Praise for David Zurick's Books

"A profoundly self-reflective meditation ... should be required reading for all conscientious adventurers."

— *Outside Magazine* (for *Errant Journeys*)

"A thoughtful and sobering account of a personal inner journey."

— *Small Town* (for *Errant Journeys*)

"I cannot emphasize enough how wonderful David Zurick's pictures are — evocative, alive, luminous, and masterful. He makes the sacred come alive."

— Jonathon Mirsky in *High Peaks Pure Earth* (for *Land of Pure Vision*)

"Those who are embarking on their own nascent journeys will gain insight through Zurick's images."

— Lindsay Skog in *Himalaya* (for *Land of Pure Vision*)

"Concise and well-written, this will interest naturalists ... and readers of natural history."

— Carolyn Alexander in *Library Journal* (for *Hawaii Naturally*)

"The enticing descriptions are enough to invite mad 'stuff it' phone calls to bosses."

— *Whole Earth Catalog* (for *Hawaii Naturally*)

"This isn't just a pretty picture book, it's a truly fascinating read."

— Rebecca Stephens in *Geographical* (Royal Geographical Society) (for *Illustrated Atlas of the Himalaya*)

"Brilliant . . . with clarity and passion."

— Regan McMahon in *San Francisco Chronicle* (for *Illustrated Atlas of the Himalaya*)

"Everything you could ever want to know, through words, maps, and photographs, of the Himalaya."

— Mary Ann Gwinn in *Seattle Times* (for *Illustrated Atlas of the Himalaya*)

"Simply gorgeous . . . as beautiful as it is informative. Manages to be all things to all readers."

— *Missoulian Sunday* (for *Illustrated Atlas of the Himalaya*)

MORNING COFFEE AT THE GOLDFISH POND

MORNING COFFEE AT THE GOLDFISH POND

Seeing a World in the Garden

David Zurick

Shanti Arts Publishing

Brunswick, Maine

MORNING COFFEE AT THE GOLDFISH POND
Seeing a World in the Garden

Published by Shanti Arts Publishing
Cover and interior design by Shanti Arts Designs

Shanti Arts LLC
Brunswick, Maine
www.shantiarts.com

Printed in the United States of America

First Edition

In this book, the author refers to the goldfish pond as "his" goldfish pond, the gardens as "his" gardens, and so forth. David Zurick lives with his wife, Jennifer, in Wolf Gap Holler, and they work together in the gardens. It is "their" home, with its gardens and goldfish pond, and none would be as lovely as they are, or even exist at all, if they were only David's.

Image of waterlily used throughout: iStock.com/Mellok; image of fish on cover courtesy of Freshwater and Marine Image Bank, University of Washington, Seattle, Washington, Wikimedia Commons.

George Strange, "Screen Test" (2011). Unpublished poem. Used with permission of the author.

Nanao Sakaki, "A Love Letter," in *Break the Mirror: The Poems of Nanao Sakaki* (Nobleboro, Maine: Blackberry Books, 1996). Used with permission of the publisher.

ISBN: 978-1-947067-02-8 (softcover)
ISBN: 978-1-947067-03-5 (digital)

LCCN: 2017950158

For my Wolf Gap friends and neighbors

and in memory of Julia (Jay) Hensley,
who found solace at the goldfish pond
during her final months of life.

Contents

The garden of the world has no limits, except in your mind.

—Rumi, thirteenth-century Persian Sunni Muslim poet, jurist, Islamic scholar, theologian, and Sufi mystic

CHAPTER ONE

It's at the point of turning onto Wolf Gap Road where George and I feel as though we've left behind the rest of the world. We are "sliding into home," and this always provides us with a considerable sense of relief. It is as though we'd gain admittance to a place.

Sliding into Home

Wet season streams flowing through the Appalachian foothills have carved Wolf Gap Holler into a tightly-wrapped little valley. Its cleft terrain isolated the frontier settlers and offers today's residents a private and protected place to live. Early on, the single lane leading into Wolf Gap was dirt and then gravel, but now it's paved with a thin layer of asphalt; people had begun to complain about the dust. Most of us have known one another for many decades and neighborly waves of greeting are commonplace in the valley. The elderly man who lived in a small house at the foot of Wolf Gap Holler, where it opens onto the Red Lick Valley, played a gatekeeper role in the holler for over half a century. Matt always wore a cowboy hat and pistol, and would linger by the side of the road inquiring of people headed into the valley what business they had there. It kept the miscreants away. He died a few years ago, and his widow Ruth lives quietly in the old house beside a failing barn amid green pastures and the gathering tributaries of Still House Creek.

My neighbor George and I often ponder our valley's history when we sit beside my goldfish pond in the garden and drink coffee. This local history is so much a part of our own life, the source of memories that make us into who we are. We've always said that once we leave Red Lick Valley and head into Wolf Gap Holler, we feel as though we are "sliding into home." The welcoming feeling was there when the road was dirt, and it still is, even though the visceral effect has diminished a bit with the new tarmac surface. Our day-to-day work takes us into the nearby towns—the outfield, a busy world—but at the end of the day, our commute brings us back to Wolf Gap through places of increasing beauty and diminishing activity. Close to home, the landscape opens onto farms and prominent ridges known as the "knobs." Local people call the hills "knobs" because they look like giant door knobs stuck to the valley floor. Geologists more precisely refer to them as monadnocks. They run up and down picturesque Red Lick Valley as knobby extrusions of the planet's crust, hinting at the calcareous world of caverns, sinkholes, and subterranean streams that lurk just beneath the surface. For this, Kentucky is justly famous.

Along much of its stretch, the Red Lick Road is flanked by weathered tobacco barns and open pasture, and the turn into my valley is marked by a dented, green-and-white sign announcing "Wolf Gap." It's hardly a proclamation. The marker sits on a pole at a slightly bent angle like a tipsy sailor saluting the passing traffic. It's easy to miss, and anyway, the road sign is misleading because a gap normally joins two sides of a mountain while Wolf Gap is a dead-end lane. Partly due to this topographic misnomer, but mainly because we live there, it's at the point of turning onto Wolf Gap Road where George and I feel as though we've left behind the rest of the world. We are "sliding into home," and this always provides us with a considerable sense of relief. It is as though we'd gain admittance to a place where they will never find us, although it's hard to say just who "they" might be, or why anyone would be looking for either of us in the first place—after all, we're not criminals or celebrities.

George has lived in Wolf Gap Holler longer than I, so his

memories of it extend further back in time than my own. He was one of the modern settlers of the valley, a member of a tribe of hippies who arrived in the valley in the early 1970s. At that time, Wolf Gap was just opening up to outsiders. It always had been a place on the outskirts of American society, with an iconoclastic history of pioneer families, slave traders, moonshiners, and tobacco farmers, most of whom basically wanted to be left alone. George and his companions proved not to be so different in disposition from their Appalachian predecessors. When we sit by my goldfish pond and recollect life in Wolf Gap Holler, it is with an eye toward its most recent history: the modern decades of residency by the back-to-land types and not the earlier generations of frontiersmen with whom we share no natal affinity and thus for whose descendants we'll always remain newcomers and outside the pale of local mythology.

George once explained to me how, early on, he and his outcaste friends had to prove themselves worthy to a segment of the Red Lick populace. That famously occurred one day when a long-haired kid named Gary, hailing from Pennsylvania, greeted a bunch of visiting Red Lick trouble-makers with a volley of rifle shots. The local inhabitants understood the message, left his premises, and peace has pretty much prevailed since in our little valley. Guns are popular in my part of the country, but only once to my knowledge has someone been shot in Wolf Gap. People target practice quite a lot, but not normally early in the day, and so, outside of the deer-hunting season, gunshots rarely disturb our morning chats by the goldfish pond. If George and I were to hear rifle shots at that time in the morning, we'd probably think something was seriously up. One of the last times I saw Matt before he died, he was wearing nothing but skivvies and holding a shotgun close to his chest, standing on the rickety front porch of his house with an oxygen tube dangling from his nose, peering across the pasture, listening for coyotes that he feared were headed toward his chicken coop. He was ninety years old and suffered respiratory and other health problems.

The term "hippies" has been used to describe the young people who moved into Wolf Gap Holler during the early 1970s. It's a loaded term that I generally dislike because it does a poor job of explaining the people it means to describe, but that is what many of the Red Lick residents called the newcomers in the early years. I don't often encounter the word nowadays, but not too long ago I overheard someone describe Wolf Gap as "Granola Road." Stereotypes die hard. The Wolf Gap population has become diverse over the years. We live amid a mix of suburban-type houses, yurts, farmsteads, double-wide trailers, hand-crafted cottages, wood shops, or straw-bale domes. Among us are teachers, potters, carpenters, farmers and factory-workers, artists, pensioners, and disability-check collectors. We come from states like Kentucky, Georgia, Michigan, Ohio, Vermont, and Pennsylvania. We generally get along fine, although schisms exist in Wolf Gap society. It's not like we live harmoniously in the valley as brothers and sisters residing in a love commune. We're not living in the middle of a social experiment. It's nothing like that at all. We're neighbors. The thing that brings most of us together is a mutual regard for the natural beauty of the place where we live and a desire to live in kind regard for one another. We enjoy, too, the advantages of being at a minor distance from popular American culture, which for some of us has taken on a disquieting quality, and within a lovely forested cove where there is sufficient space that we need not worry too much about bumping into one another. We like our scenery and elbow room as much as companionship. It's the ease that comes from living within a small fold of Earth, far from the jostling crowds, that also makes it seem to George and me when we turn onto Wolf Gap Road that we are sliding into home.

As time goes by and my residency in Wolf Gap Holler

lengthens, a sense of being at home in the valley intensifies, and the measure of contentment I feel when I return to it increases proportionately. This is amplified the harder I work in my gardens and reinforced each time I sit at my goldfish pond, which is to say that such an expenditure of time and energy heightens my connection to the place where I live. It's a natural response, what most people will feel if they give it a chance, and thus, quite frankly, is cause for hope for humankind. Still, it's just been a couple of decades that I've lived in Wolf Gap Holler, and my sense of belonging to the place doesn't extend far beyond the small community where I live. The popular society around me feels alien, and the surrounding countryside, pleasant as it is, remains obfuscate from my perch beside the goldfish pond in the garden. Having grown up on Lake Huron in Michigan and then lived in Hawaii for many years, I'm always looking, still, for a big body of water somewhere beyond the next rise of hills. I know it's not there, and that realization leaves me unsettled, as if my interior geographical puzzle is missing a piece. Maybe the lack of a shoreline explains the goldfish pond. Another problem for me living in rural Kentucky, though, is that I have no birthright to the place. Without that, I'll never fit into local society beyond my chosen acquaintances, most of whom also hail from someplace else. Even among my Wolf Gap neighbors, who are a mixed lot of immigrants and locals, there exists a sorting out of people according to social distance and natal allegiance. What does it mean, then, for someone to feel connected to a place they have occupied for millennia and, moreover, that holds a sacred meaning for them? What kind of home plate is that?

For several thousand years prior to the arrival of white settlers, Wolf Gap Holler was home ground to Amerindian peoples: the Adena Culture and Hopewell tribes during the Woodland Period (100 B.C.E.), the Fort Ancient Culture beginning about 1000 C.E., and more recently, hunters and foragers of the Southern

Cherokee Nation before the U.S. Government sent them packing to eastern Oklahoma in 1835 on the "Trail of Tears." There's not much of a record of the early inhabitants in the local landscape; a few ceremonial mounds lie scattered in the countryside if a person knows where to look, but most native sites were obliterated over the years as the land was cleared and farmed by white settlers. Artifacts like spear points and arrowheads occasionally turn up in the soil during the spring plowing season, but little else remains of a material record of the valley's original inhabitants. It would be hard to know that for native people the Red Lick Valley once was sacred ground. A ridgeline that overlooks Red Lick a few miles to the west of Wolf Gap Holler is known as Indian Fort Mountain. It was a refuge of the early Hopewell tribes, reputed to have been a ceremonial place, but is now a popular hiking area.

I can imagine, though, how the land around me might have been apprehended in spiritual ways by ancient people. In part, that's because I have come to see the place where I live — the immediate environs of my house, gardens, and goldfish pond, and to a certain extent, the world-at-large — in a divine way because I spend a lot of time in places where people maintain long-standing cultural rights over their ancestral lands and regard their holdings in a sacred light. I find such places to be instructive because otherwise I live in the United States, which is fast becoming an antithesis of such a holy comprehension of the world, and in a society that long ago rejected native ways of apprehending the world, and which, nowadays, seems driven by such a blinding lust for growth and consumption that, gone unscrutinized, may only come at the loss of our spiritual past and the planet's ecological future. The native places I visit remind me that there exists another way of knowing the world. One of the most enlightening spots for me in this regard is the Tibet Plateau, where the very idea of sacred geography lies at the heart of its culture.

Although Tibet is located around the globe from Wolf Gap

Holler, there are some curious connections between Kentucky and Tibet apart from my own perambulations in that part of the world. The Native American tribes trace their ancestry to Siberia and Central Asia, including Tibet. The archaeological record shows that the earliest migrations began about 30,000 years ago, when travel across the Bering Strait was possible, and over the ensuing millennia resulted in the full-blown colonization of North America, including the Red Lick Valley. Over time, the wanderers were no longer Asians, having distanced themselves by such a huge gulf of time and space from their home ground dwellings, and evolved into the New World peoples. Attempts to tangibly link the aboriginal Americans to their Asian forebears have led to all sorts of scientific hypotheses and more than a few fanciful conjectures. Genes and mythology tell the story. The clincher for me, though, occurred when I stumbled upon a dreamcatcher inside a yak herder's tent in eastern Tibet, recognizing it from ones I've seen in Kentucky and elsewhere. I first observed the talisman hanging from a pole as I settled down for the night with a family of nomads, all sprawled together atop animal hides and covered by a prickly wool blanket. The totem dangled brightly above, catching the last bit of light emitted by the embers of the hearth fire.

A dreamcatcher is a tiny net of woven sinew adorned with feathers, beads, or shells that traditionally is hung above sleeping quarters to keep away bad dreams and evil spirits—to protect believers during the dark hours of the night. Some historians contend they originated among the Ojibwa tribes of the Upper Great Lakes region, but it's hard to say because they are so widely distributed in North America as a cultural identifier for many native people. So what was one doing with a yak herder on a remote plateau north of the Himalaya Mountains? When I saw the dreamcatcher in Tibet, I knew right away it had the same design as the North American versions. And when I asked my host about it, he explained the talisman warded away bad spirits, and that all the nomadic families thereabouts possessed one. Such a specific article replicated among cultures separated by many thousands of years and half the circumference of the planet...what are the odds of that?

Anthropologists are quick to point out that certain features of a culture might be mimicked by others when they play a similar function in society. It follows that specific objects (for example, a tool or a body adornment, and maybe a dreamcatcher) might develop in disparate localities along shared designs because they enjoy the same basic meaning for different people. Culture diffuses across great distances with trade and travel, and so widely dispersed societies may share fundamental qualities of certain objects because they have been borrowed. I was aware of such theories in the herder's tent, but I also thought there had to be a more specific and reasonable explanation for the dreamcatcher in both Tibet and Kentucky. I know I didn't carry one from Wolf Gap across the Himalaya to the Tibet plateau. Maybe dreamcatchers were taken to Tibet by other romantic types visiting from America? Or did Tibetans see pictures of them in books or magazines and copy the style at home? Or was it the other way around? Such scenarios are highly unlikely. When I asked him about it, the herder said the dreamcatcher, known in Tibet as a *namka*, was "old-before-time." A Cherokee friend of mine on Red Lick Road, meanwhile, explained to me how the dreamcatcher had been with his people for as long as anyone could remember. It remained difficult for me to understand, though, how something as ephemeral as a tiny symbolic weaving could cross all that history and geography.

My grasp of how cultures might be connected was furthered by learning that a group of Navajo elders once visited Tibet and immediately began to cry upon arrival. They explained that they felt such a deep empathy at having "returned home" that it moved them to tears. The red-rock terrain and sunny climate, sheep grazing on wide-open pastures, were strikingly familiar to them—a mirror of the Arizona landscape. I could imagine them being at home in a Lhasa marketplace, among the bins of silver and turquoise ornaments, pottery, and piles of tribal textiles so alike in color and geometric pattern to those found on the Navajo reservations. The Navajo visitors also commented on Tibetan architecture: the teepee shape of the nomads' tents as well as the more substantial adobe villages that resembled their pueblos on the Kaibab Plateau. With tiny windows and doors, and thick mud walls, the structures in both places are designed to keep their inhabitants warm in winter and to safeguard them from the harsh glare of the summer sun. More than anything, though, the Navajo elders recalled an emotional sensation of being at home in Tibet. They sensed that the spirits that dwell in the Tibetan landscape are the same ones they knew at home in Arizona.

In Tibet, the land, itself, is believed to be sacred ground, and some places are deemed to have special powers due to their geography or cultural history: a summit, stream confluence, forest grove, or the birthplace of a saint. First the shamans and animists, and then the Buddhists, built their shrines and monasteries at the most auspicious places. They include the eponymous earth-taming temples, designed for nothing less than to modulate the cosmic energy of the planet. Some places are believed to be so holy as to constitute a topographic coupling of Heaven and Earth, the axis mundi: a mystical lightening rod to the Empyrean. Scattered among Tibet's manifestly sacred places are hidden valleys called *beyuls*, the holy of holies, believed to contain ancient meditation treatises left behind long ago by wandering sages, mystics, and bodhisattvas. Such places are to be discovered not by cartographic coordinates, which may be ambiguous or nonexistent, but rather by the spiritual preparations of those who search for them: people traditionally called *tertons* or "treasure-seekers." In all such quests, a *beyul* is not so much a place to be found as one revealed, but then only to a person who is spiritually prepared to recognize it.

Tibetan pilgrims often travel by foot in full-length body prostrations and may spend a year at a time on legendary journeys to the holy places, accompanied by a spouse or guardian pulling a rickety cart filled with their worldly goods. I've met pilgrim caravans journeying a thousand kilometers or more by bowing and scraping their way across some of Earth's most forbidding terrain. Upon reaching their destination—a temple, cave, or mountain spring—the pilgrims would circumambulate it three times in a clockwise direction, spinning prayer wheels, chanting, and fingering rosary beads, pinning their hopes of salvation to the esoteric conjunction of faith, ritual, and geography.

Back in Kentucky, George and I drive our aging pickup trucks into Wolf Gap Holler with no particular inclination to making a genuflecting entry into it. We're not Tibetan pilgrims on a spiritual quest, but modern Americans leading ordinary lives. Still, when I arrive in Wolf Gap Holler, and especially when I am seated beside my goldfish pond, I can open my heart to the sanctity that resides there and find purpose in the natural qualities of the place, a deep-dwelling realization that George and I simply describe as sliding into home. The Tibetan pilgrims have never set foot inside an American ballpark, but to see them in the midst of their epic prostrations is to glimpse a person literally diving head first into the home plate of Tibet, covered in a cloud of sacred Earth dust, running a baseline from sacred place to sacred place, from faith and ritual to the hope of spiritual redemption. Talk about hitting a home run!

The sacred places I've visited in Tibet have a palpable atmosphere. Sometimes, I enter one of the old temples and my skin begins to jump with electricity, as if I was inside a prayer bell that had just been vigorously rung three times by a levitating lama. I love those kinds of places. The buildings are made of wood and stone, filled with burning incense, yak butter lamps, and musty carpets, with walls encrusted in richly colored murals depicting demons, warriors, or delicately wrought saints, rooms resounding in a cacophony of chants, drumbeats, and clashing cymbals. Such places hold a mysterious history that lies beyond my own experiences. The dizzying sensations I feel inside a monastery are to the point of mental distraction, opening me to the possibility of revelation because, oddly, this sort of sensory dislocation empties a person's mind—a necessary prerequisite for meditation—and awakens me to the prospect of a new reality. I eventually came to see in the temples of Tibet a ritualized loci of devotion formed by the intersecting lines of faith. The pilgrims' journeys cover the plateau in a mythic geometry that is as purposeful, true, and necessary to a Tibetan's comprehension of their sacred world as are the lines of latitude and longitude on a map to a cartographer's reckoning of the physical contours of the planet.

When I return to Wolf Gap from Tibet, I always carry with me a trace of this fragile notion of sacred geography, like the scent of a far-off place: a rare and valuable souvenir. Rather than let it simply drift away as a fleeting memory, I try to cultivate its presence in my home, among the gardens and forest, so that it can linger and have a sustaining influence on my life. The benefit of this is great. For all the appeal of a Tibetan monastery, ripe with age and the mutterings of monks, a similar reverence can be had near to home, at my goldfish pond in the garden, where I sit amid the efflorescence of water lilies and listen to the bass notes of the croaking bullfrogs—which sound mysteriously like chanting Tibetan priests—smell the magnolia blossoms, and watch mesmerized as colorful fish school in the water while butterflies and hummingbirds cavort overhead. Seated in the garden, my mind empties just as readily as if I were sitting zazen in a Buddhist monastery on the opposite side of the world. Over time, the goldfish pond has become a shrine in my own home-ground temple, a tiny ecosystem encompassing the human spirit and natural world in a harmonizing way. It is a pleasant place to enjoy a cup of coffee in the soft morning light with my neighbor George.

The Tibetan people have been practicing their faith for many centuries inside their old temples, sanctifying them by ritual and careful observance, and, by so doing, making them holy. Their spiritual outlook extends far beyond the confines of a singular monastery. It stretches across the vastness of the Tibetan plateau, the largest patch of high ground on Earth. Their holy ground includes the planet's tallest mountains and deepest gorges, nestles deep into caves and forest glens, flows through glacier-fed streams and pools within holy lakes, is emboldened by colorful prayer flags flapping loosely in the icy wind, and is witnessed through ages by lichen-covered mani stones containing chiseled bits of scripture, by spinning prayer wheels, and by a network of trails worn smooth by the feet of passing pilgrims. From this ancient landscape emerges an idea of what the Tibetans might be thinking when they look out upon the holy *mise-en-scène* of their home ground: an entire world that is sacred.

When I visit the Buddhist monasteries in the Himalaya, it's not as an especially devout person learned in canonical ways. I can't seem to remember the lineages of the Tibetan saints or important passages of scripture. I find wisdom, though, in the tantric teachings, and I especially enjoy being in one of the old sacred places in Tibet; they always seem to provide me with a holy moment. At Kumbum monastery in central Tibet, I discovered a fantastic temple painted in the creamy white colors of the snow-capped peaks. The building was decorated with exquisitely detailed murals and four sets of Buddha's protective "all-seeing eyes" painted onto its uppermost balcony. The boldly drawn gaze looked out upon the mountains that surround the Nyang Chu valley. Dozens of tiny portals on each of the temple's nine levels, or *lhakangs*, led into dark chambers filled with tiny golden statues and paintings of deities and demons (*kumbum* means 100,000 images). The spiral architecture of the Kumbum monastery formed a three-dimensional mandala—Buddhism's renowned symbolic map of the world, whose iconic elements suggest we all move through multiple incarnations toward greater awareness and compassion. The mandala is a powerful symbol throughout Tibet, and the Kumbum monastery is one of its most striking architectural manifestations.

I climbed higher in the temple and entered a succession of dim chapels. Stepping out from each and back into the full sun, repeating the movement time and again, I slowly made my way to the apex of the hemispheric structure. The sensation finally was one of a bright light being rapidly switched on and off inside my head. The tiny murals set deep within the dark recesses of the temple merged into a single image, the wisp of an idea that floated out of reach when I sought to focus my thoughts upon it—stigmata without form, just vague and colorful burnings inside my mind. The passageway spiraled through multiple tiers of circular balconies, and I was enveloped in the heart-thumping sensation that perhaps I really was inside a mandala; if not, then a wildly-spinning gyroscope set loose on the sacred terreplein of Tibet. I paused to rest when I attained the summit of the temple and let my mind slowly unravel at the marvelous sight of icy mountains shimmering in the distance. A deep calm passed through me and the breath lost to altitude and exertions gradually returned to my body.

The Kumbum monastery is one of many such places in Tibet. What sets it apart from others on the plateau is the fact that it remains in very good condition. The vast majority of Tibet's monasteries were destroyed by the Chinese army, first when it invaded Tibet in 1950 and later in the 1960s, during the

Cultural Revolution, when religious sites throughout China fell to the blunt axe of Maoist ideology. The uncomprehending Chinese soldiers, spurred on by their wanton generals, mercilessly shelled the temples in Tibet and used them for target practice. They not only blasted the religious structures but plundered their wealth of golden statues, massacred tens of thousands of monks, obliterated entire monastic universities and medical colleges, ripped down prayer flags, and burned huge piles of scripture in village squares. The sacred thread binding the Tibetans to their homeland, spun over so many centuries of faith, was horribly frayed by the Chinese; their home plate was stolen from them.

I think about all that whenever I visit Tibet. And I try to imagine not just the deep empathy the Tibetans have for their place in the world, but also what it must mean for them to know that it has been ruthlessly severed and most probably will never be restored. I also think about it sometimes when I'm in Wolf Gap Holler, seated beside my goldfish pond, connected to my home in ways that only grow stronger with the passage of time—and still the time I have spent there is so short—and what I might feel if it all were to be destroyed by some thoughtless and violent interloper? I would feel like the rug had been pulled out from under me. Like a cold wind was blowing through a hole in my heart. I know I'm fortunate to live among the soft hills of the Red Lick Valley, nestled in a forest and surrounded by mainly thoughtful and courteous neighbors, at a great distance it would seem from outside invaders, and with little to fear of the tragic events that unfold in places like Tibet, where the depths of despair at having lost their place in the world lead some people to acts of self-immolation and others to lives of prayer and compassion. The minor alienation I feel amid the larger society within which I live, and of which I have no personal history beyond that of my neighbors and friends, gradually fades as I enter Wolf Gap Holler and finally dissolves into the gentle surface of my goldfish pond. ❖

CHAPTER TWO

A large clump of root balls had upturned where the pine trees fell atop the structure, lifting the earth to expose the substrata of red clay soil immediately beneath its surface. The hole in the ground looked like a bloody gash in the planet's lithosphere. I eventually made my goldfish pond in this spot, and it's now a centerpiece of the gardens, a place of solace sculpted in the wake of a winter storm.

A Storm and a Garden

A late season storm dropped two feet of wet snow on Wolf Gap Holler, felling trees in the forest, downing power lines, and closing roads. The weight of the snow settling onto the pines and cedars broke them at the trunks or caused them to bend in great arcs that spanned the woodland trails, driveways, and roads out of the valley. I lay awake much of the night of the storm, listening to the groans and explosions made by the tortured trees in the woods, waiting for one to crash down upon my house. I awoke to a scene of tumultuous beauty and devastation. On the one hand, the snowfall on the distant hills made them sparkle in the morning light so they had a painterly effect. Closer in, the forest was a flattened and tangled mess, as if a cyclone had decided to settle down in the night and build a nest in the valley. Every so often, a loud pop and ripping noise would break the hoary silence of the valley as yet another tree yielded to its frozen encumbrance.

A nearby grove of conifers was a dangerous knot of toppled tree trunks, smashed limbs, and gnarly upturned roots. Eight large pines had fallen in a crisscross fashion onto my studio's roof, warping the building's frame with their combined weight so that I couldn't open the door. Cedar trees in white greatcoats of winter huddled close to the ground, still alive but looking forlorn and exhausted from the effort of standing upright against the snow that had accumulated in the night. They drooped in the forest like an army of fallen Cossacks. It appeared that I was no longer living in a cozy cabin in the woods, but rather amid a damaged and alien-looking landscape. Peering out upon the devastation, I could imagine what it might be like for a person to survive a tornado, step outside their home, and discover their neighborhood is in ruins. My situation was not so dire, but, still, I learned more about the capriciousness of natural calamity and the disequilibrium that might visit those who survive one. The trees, meanwhile, kept popping and ripping in the forest around me as they gave way to their blankets of snow. With amazement, I wondered at the violence the delicate snowflakes are capable of when enough of them are piled onto a single spot.

I walked around my property and took its measure. The studio looked a wreck. Hoping to relieve it of some of their crushing force, I climbed with a chainsaw atop the roof of the structure and began cutting away the limbs of the fallen trees. The overhanging drifts of snow made it difficult to find the roofline so I treated its edge like I would the cornice of a snow-capped mountain and stayed away from it. When I first touched my saw to the trunk of one of those downed trees, it exploded like a time bomb, as if it had been waiting for me like one of those roadside detonations we hear about in the news that kill so many people in war zones these days. It knocked me backward but not off the roof. The tree was twisted out of any natural symmetry, locked in a contortion that would make a Romanian gymnast blush, and snapped dangerously when I released it from the tension that held it together. I was clearly out of my league amid this wicked snarl of growling limbs and thick-set trunks.

I put the chainsaw down, paused to consider my next move, and watched as two of my neighbors post-holed across a snow-drifted pasture toward my house. They had seen me working on the roof and wondered about the damage to it. Mark climbed onto the building and immediately got busy with my saw. He was young and strong. Clumps of snow fell onto his dreadlocks from overhanging branches, so that he soon looked like a Rastafarian iceman far removed from a green and warm island. His exhalations were puffs of steam chugging right alongside the screaming chainsaw. The building groaned when Mark cut the last tree from it, then it shifted a bit and settled back into itself. I tried the studio door again and opened it easily—just like that—as if nothing untoward had happened to the structure. My neighbors left to survey the other neighbors' damaged properties.

It took five days to clear my lane and the Wolf Gap road out to Red Lick Valley. Everyone pitched in to help one another out. The trees kept falling behind us as we cleared them away in front. At the end of a day, looking back upon a newly-opened stretch of road and feeling good about it, we knew the following morning would find it blocked once again. The trees just kept dropping. The strongest ones succumbed slower to the force of the snow, but it seemed as though at some point all the trees in the forest might tumble like dominoes, and I would be entirely surrounded by bald ridges and a scarred hillside devoid of any lasting verdure. Trapped in our valley for the duration of the storm's aftermath, we cooked food on woodstoves and shared meals in the evenings. The normal routines of career and commute, the conveniences of electronic devices and communications, all the stuff of modern life, were suspended by the closed roads and downed power lines. It was a time instead for hard work, oil lamps, potlucks, and home fires, and greeting friends and neighbors as we all slogged along broken trails in the snowy woods with pots of food, chainsaws, and tins of gasoline and oil.

My life took an ordinary turn once the road out to Red Lick was opened. I went back to my job, and the electricity returned to my house. It was another month of steady work, though, before

I managed to clear the yard around my home of the botanic debris. And I labored another five months in the woods to make space in the canopy for the understory of spindly saplings. The young trees had been overcrowded by the towering pines, but with the removal of storm wreckage, they would now have the room and light to grow into great specimens. My interventions in this regard were decried by some persons who viewed the snowstorm as naturally resetting the ecological clock in the woods, but from my perspective, I'm part of that sylvan timepiece, and I thought I could move things along toward a quicker recovery by making room in the woods so the young trees might reach maturity in my lifetime.

Much of the initial salvage work centered on clearing the space around my studio building. A large clump of root balls had upturned where the pine trees fell atop the structure, lifting the earth to expose the substrata of red clay soil immediately beneath its surface. The hole in the ground looked like a bloody gash in the planet's lithosphere. I eventually made my goldfish pond in this spot, and it's now a centerpiece of the gardens, a place of solace sculpted in the wake of a winter storm. In this manner, I followed the path of people everywhere who turn adversity on its head. It's a common experience of humankind—this making lemonade of lemons—but in India, an entire religion is built upon the notion that in destruction lie the seeds of creation. For Hindus, karmic theory governs all human consciousness and allows people to enjoin their lives in great cosmic swirls of destruction and creation.

I often travel to India and have found that nowhere is this conflation of life and death more evident than along the banks of the Ganges River in the holy city of Varanasi, where on a daily basis going back at least two thousand years, people have burned the bodies of the recently deceased in the hopes of *moksha*, the release of a person's soul from the endless cycles of worldly incarnation. It is believed that in this final act of consummation by fire, a person's body releases its spirit to rise again in another life, or in acts of pure salvation, dissolve straight away into eternal union with the Divine. I know it's an improbable stretch of the imagination to hitch my goldfish pond to that team of celestial horses. But so it is. When a winter storm moved through my valley and knocked down the trees around my home, I was left with a clay depression in the ground and the opportunity to make from it an earthen basin of water, to fill it with flowering plants and colorful fish, and to find there a tiny world of constant wonder.

The spiritual center of Varanasi is the Manikarnika ghat, an ancient, worn rock staircase descending into the Ganges River where bodies are cremated so their ashes might be directly dispatched into the currents of the sacred river, thus deporting a person's soul upon its epic transmigration. I sat on the steps of the ghat one winter morning and watched the fog-wreathed river slowly come to life. Pilgrims descended to the water's edge, disrobed, entered the frigid shallows, and prayed with uplifted palms to the rising sun. Small wooden rowboats came and went in the thick dawn mist, their ghostly passage marked by the quiet slap of paddles dipping into the water. Deep-throated mantras and piercing sounds of temple bells emanated from the nearby structures. Monkeys scampered among the stone-and-plaster buildings, behaving like masters of parkour as they brachiated wildly across the ancient rooftops in their barking and hissing pursuit of one another. Close by, elderly, stooped penitents wearing woolen shawls and tightly wound head scarves shuffled past, using canes, mumbling, carrying small offerings of flowers and incense in their old, wrinkled hands; not a one noticed my presence, so intent were they upon their deliverances and well-accustomed to all manner of people sitting beside the great river.

My attention was diverted by the sound of mournful singing. Feet thudded in union against the stone pathway, growing louder as a small cortège of shaven-headed men carrying a bundled body appeared at the top of the ghat. They wore white sarongs tied at the waist, and sacred threads of red cotton were draped against their naked chests. Plastered turmeric etched the men's foreheads with the yellow markings of Shiva's trident. The men walked purposively with their load toward the river. Next came a small cluster of women draped in saris with drawn veils hiding dolorous faces, chanting sorrowful mantras through the fabric or singing softly with tears in their eyes. The funerary procession moved close to the river and halted at a carefully stacked pyre of wood upon which the men placed the body with feet pointing symbolically south toward Yama, the Lord of Death. After making some minor adjustments to the shroud, fussing with some loose strands of fabric, the eldest son of the deceased person slowly circled the pyre three times in a counter-clockwise fashion, praying softly all the while and sprinkling the corpse with river water. He stopped and tenderly inserted a few drops of the sacred Ganges directly into the shrunken mouth. Then he stooped and ignited a tinder fire of cow dung and ghee, and placed the flame gently onto the kindling beneath the body of his mother. Within minutes the pyre was consumed by fire. The mourners left without a backward glance, leaving the still burning body to be managed by a ghat attendant who, for a small fee, tended the flaming corpses deposited on the riverbank and thus served as boatman for the deceased across the River of Death. When the fire was fully out and the cremation complete, family members would return to the site, gather the still warm ashes, and strew them into the river.

I watched the ceremony from my perch on the stone staircase and was struck by the resoluteness of these last rites. The grief was evident, especially in the countenances of the women, but the overall atmosphere was of acceptance rather than of anguish or despair. I got the sense that the injunctions of immateriality that lay at the core of Hindu philosophy are taken to heart in these cremations, which are so commonplace in Varanasi that the city's history is built upon them. Outside the circle of death, beyond the burning pyres, the Ganges riverbank was a riot of color and exotic ritual. Persons from throughout India drifted past me in thick ribbons of humanity, streaming toward the river to complete their arcane ablutions before returning again to the labyrinthine folds of the ancient bazaar. The pedestrian traffic vibrated with purpose and devotion. Known among the Vedic scriptures as Kashi, or the "City of Light," Varanasi has always been a place where people simultaneously celebrate life and death in the hope of gaining release from a Karmic embrace. This day was no different.

Saffron-robed priests ambled along the steps of the ghat, collecting alms from penitents and pleading for charity. Some of the clergy approached and asked me to purchase a threaded bracelet for luck or to make a donation, but to what effect? Infirmed beggars, some blind and others with missing limbs or lost minds, lined the pathways in squatting rows of upheld palms and blank, wide-eyed stares. Holy men and charlatans, wearing their hair in knee-length dreadlocks and adorned in silver jewelry and painted faces, sat on felt blankets telling fortunes and smoking hashish out of tubular clay chillums. Street kids swam and frolicked in the horribly polluted river amid the pilgrims, priests, hucksters, and ever-present signs of death, as if they hadn't a care in the world. A band of sweepers passed silently, moving through the crowd like a swarm of locusts, raising Ganges sediment and pyre ashes with each swirl of their straw brooms. The sun rose in the sky and dissolved the veil of mist on the river, allowing the far side of it to come into view. I watched desultory cattle come into focus as they grazed somnolently amid the crumbling structures on the opposite riverbank. With the incandescence of the sun came also the fragrance of burning sandalwood and juniper, the sweet stench of charred human flesh, and the mixed odors of the city: diesel exhaust, rotting garbage, industry, and the scent of hot curry.

A cremation is considered to be a beginning, not the end, of a person's life, and when finally the ashes of a person enter the Ganges, it is not in the way of crossing the River Styx as in Greek mythology, which leads to a heinous underworld, but rather of moving across a threshold from one life to another into a more hopeful future. Looking down the length of the Ganges River, I observed other ghats where similar rituals were taking place. Smoke from the numerous cremation fires became voluminous and cast a yellow pall over the scene, replacing the mist that had dissolved in the gathering sunlight. As the morning progressed, the ghats filled not just with devout supplicants or family members attending the cremations of loved ones, but with hordes of gawking tourists, vendors peddling hot tea, tradesmen, idlers, and curious onlookers who, like me, seemed intent upon gaining from the spectacle a new outlook on the subtly intertwined matters of life and death.

While the Ganges River in Varanasi is one India's holiest places, it is not at all like a Christian church, a synagogue, or Islamic mosque, where an inviolate sense of sanctity defines the religious space. Pilgrims come to the river to purify themselves in water that I found to be filthy almost beyond comprehension. The noise and bustle of the Varanasi quay would prohibit the propriety found at most other religious places in the world. Profanity mixed freely with the sacred Ganges rituals until one almost became the reflection of the other. It was an intoxicating and bewildering assault on the senses. And while the river is given over mainly to spiritual pursuits, however blunted they may appear to the uninitiated, the twisted lanes of the city beyond its banks is a teeming marketplace of deceit, duplicity, and dubious goods. Merchants freely mingle with pilgrims, thieves, and beggars, until all seems to be one. The Byzantine alleyways harbored shops that overflowed with religious goods, spices, perfumed oils and incense, silks, and antiquities. Most of the items were of questionable provenance and none could be had without merciless bargaining and name-calling.

As the years pass, in looking back upon my experiences there, I'm not aware that my thoughts about Varanasi are settled in any definite way. I visit the city whenever I have the chance and continue to find it a grand, eye-opening celebration of life. It also is a model of how profligate a society can become and still be considered holy, a lesson in how life is tempered by the looming presence of death. But I find now the possibility of as much supernal repose seated on the bench beside my goldfish pond in Wolf Gap Holler as when I'm sitting at dawn in the Ganges mist and observing the arcane rituals along the Varanasi riverbank. Such a realization is instructive, for it provides an example of how I am moving away from the postures of my youth. When I first went to India, a place like Varanasi held intense spiritual prospect for me, a place of possible redemption, but I now tend to relegate it to mainly sensual experiences, an intoxicating stew of smells, sounds, and sights that is wildly boisterous and always fantastic but not as luminous as when I first visited the city as a young man.

The most revered deity for Hindus in Varanasi is Lord Shiva. He embodies creation and destruction, and thus masks the fusion of life and death, which lies at the heart of the Ganges devotions. His visage is found everywhere in India. His name is chanted aloud in city streets. Famous temples dedicated in Shiva's honor attract millions of devotees. In a beautiful sculptural form, Shiva is with his consort Parvati in the *nataraj*, the "dance of bliss," their limbs intertwined and palms touching in a loving, spiraling embrace. The *nataraj* dance of Shiva and Parvati embodies the swirling entanglements of *samsara*, the Wheel of Life and Death, in which a person's soul manifests time and again in the cycles of reincarnation, and which symbolizes the balance between masculinity and femininity, darkness and light, and the hope of eternal reward for a well-lived life. According to Hindu scripture, a person who carefully observes the spiritual progress of their life—follows their dharma and is mindful—can escape the endless cycles of rebirth and attain a state of detached enlightenment called *nirvana*. It is with this hope that people come from throughout India to worship and bathe, and some to die, at the Varanasi ghats beside the Ganges River, a denouement that for many people is worth several lifetimes of intense spiritual devotion.

The simultaneous appraisal of loss and recovery is neither limited to India, with its promise of reincarnation and the transmigration of a soul, nor, of course, to my goldfish pond in Wolf Gap Holler, which emerged in the wake of a winter's storm, but rather is found wherever a person picks himself up and dusts herself off after a nasty fall. The news is full of it on a grand scale—war, natural disasters, genocide, tyrannical brutality, and then the behind-the-scenes stories of the survivors of such tragedies who manage to pull themselves together and make a new life. I think those kinds of success stories should be on the front page of the news rather than buried elsewhere so we might hear more about how pluck and survival is as much a part of life as death and destruction.

On December 26th, 2004, a rupture of the sea floor more than one thousand kilometers in length produced a 9.0 magnitude earthquake with an epicenter off the western coast of Sumatra in Indonesia. It caused the planet to shudder with a release of energy that was equivalent to 23,000 atomic bombs all detonated at once. Ocean water displaced by the quake sent a tsunami surging at five hundred kilometers per hour across the Indian Ocean until it slammed against the shores of India, Sri Lanka, Myanmar, Thailand, and Indonesia, and finally, nearly five thousand kilometers away, of eastern Africa. In some places, the waves crested at heights of more than thirty meters and washed away entire villages, undermining their geological foundations. Elsewhere, they surged inland for many kilometers like a powerful river. Entire islands were submerged into the ocean. Witnesses recall the sound of a jet airplane when the tsunami bore down upon them. More than 230,000 people were killed by the giant waves in a matter of minutes.

Reports of the disaster naturally centered upon its horrific aftermath, but behind the tragic headlines were countless stories of human heroism or a perceived divine intervention. A villager in India knew enough to recognize the danger posed by a receding ocean, which often occurs minutes prior to an incoming tsunami wave, and warned his countrymen to race to high ground, thus saving fifteen hundred lives. At the same time, on the holiday island of Phuket in Thailand, hundreds of foreign tourists casually ventured into Patong Bay before the oncoming tsunami, intrigued by the exposed seabed, and were caught unawares when the huge wall of water swept onto shore. An old acquaintance of mine from Hawaii who runs kayaking tours on Phuket Island recalled how on that morning he had arranged for a member of his staff to meet some clients at a wharf at Patong Bay to prepare for a tour of the coastline. The earthquake occurred at 8:00 A.M. local time, while the man was still at home eating his breakfast. He felt the ground shudder beneath his feet and watched the milk slosh in his cereal bowl. The agitated liquid triggered in his mind the connection between earthquake and tsunami, and he called his colleague on a cell phone to warn the man to immediately take his clients up a nearby hill. All of them were saved from the giant wave that soon thundered into the bay.

Some coastlines hit by the tsunami have never recovered, while in other places you would hardly know it had happened at all. People react to things and repair the places in which they live according to different capacities. Coincidentally, I was in Thailand in the days prior to the tsunami and had made plans to visit the coastal spots that were to be devastated by the tsunami: Phuket, Krabi, and Phi Phi Island. At the last minute, due to scheduling problems, I changed my itinerary and went instead to an island located on the eastern side of the Isthmus of Thailand. The big waves struck the western side. I was swimming on a reef on Christmas Day when the water was stunningly clear and filled with colorful fish and bright coral. On the following day, I went back to the same spot on the reef and found the ocean murky and clouded with floating debris. I was confused by the water's turbidity. No storm or other noticeable natural event explained it. The surface of the sea was as calm as it had been on the prior day. Unbeknownst to me, the earthquake and tsunami occurred that same morning and the seismically-generated waves, pulsating from their geological epicenter in the Andaman Sea, were pumping through the narrow Strait of Malacca and ricocheting throughout the Gulf of Thailand. I remember lamenting the poor visibility of the water, thinking it had ruined my day of ocean fun, not knowing that at the same time, on an opposite coast, the killer waves were creating unfathomable scenes of death and destruction.

Four years later, I visited many of the places that were struck by the tsunami. It was often difficult to see where the disaster had occurred. An entire town was rebuilt on Phi Phi Island with new structures designed to absorb the shock of an earthquake or the force of a tsunami wave. Emergency medical stations were established in coastal Krabi where none before had existed. Tsunami detection buoys were positioned far out into the Andaman Sea to monitor the ocean swells and predict the oncoming waves, and a hazard-warning system was installed on Phuket Island to notify people when it was time to head for high ground. Evacuation routes were demarcated so that people might know where to go in the event they had to run for their lives. I met a young Norwegian man on Phi Phi Island who related to me how he had been climbing on the cliffs above the island's Tonsai Bay near the main tourist village of Phi Phi Don on the day the tsunami had struck. Perched on a rocky ledge high above the bay, he had watched the giant wave appear on the horizon and then wash across the island isthmus, sweeping away every structure in sight and killing four thousand persons, including some of his friends who were still asleep in their tiny guesthouse. The guy had returned to the island to commemorate the passing of his friends by climbing again on its famous cliffs. He thought it might bring some closure to the fact of their deaths and help make his own life more comprehensible.

Why do I write of such events? I suppose it's to put my Wolf Gap snowstorm and the goldfish pond into some kind of perspective. People in India created a religion to help them navigate the impenetrable chasm between destruction and creation. Scientists, meanwhile, seek to prove cause-and-effect relationships that might determine the course of our lives. Some philosophers wonder if we can tell the difference. Thailand was struck by a tsunami of historical magnitude, a wave of monstrous proportions that had a devastating impact on human life, at scales that are difficult to comprehend, and while numerous lives were lost, society recovered in hopeful ways. In some instances, entire villages were moved inland onto higher ground and out of future harm's way. I don't wish to draw equivalents because there are none. In Thailand, after the epic tsunami, people buried the dead and got busy living again. In Kentucky, I made a goldfish pond from a hole in the ground in the wake of a winter snowstorm. ❖

CHAPTER THREE

Bali is known to its inhabitants as the Dawn of the World, the place on Earth where time is thought to begin. Accordingly, at first light the Balinese enter their gardens with incense in hand, pray by the pond, and pick flowers for nearby shrines. In Bali, the harmony of the world may well hang on the beauty of a goldfish pond in a garden.

Beauty and Balance in a Goldfish Pond

The frogs discovered my goldfish pond on the first day I filled it with water. The pond hadn't yet settled into the garden. Its rubber liner showed like an embarrassing exposure, spilling out of its hole and onto the bare ground, and the rocks on the bottom glistened like the glossy stones of an empty storefront aquarium. The pond looked like it had just been off-loaded from a delivery truck. Still, the frogs must have liked it when they came upon it because some of them stayed and made the pond their home. I recall it being a dry season, so maybe they'd been lying in wait all along, patiently watching me work and wondering when I might finish, and then just hopped in once I had provisioned the hole with water. Or perhaps they were passing by on their way to another destination, sensed its deliquescence, and got distracted. I'll never know for certain what those pioneer frogs were thinking when they came upon my goldfish pond. They were its first settlers. Dragonflies already hovered above the water's surface, but I can't say they had taken up residence in the pond; I remember the frogs being the first to do that. I was glad the frogs decided to stay because it gave me hope that things would work out in the end; the pond was going to be healthy. Frogs in a pond are like canaries in a coal mine: they die or leave when conditions aren't just right for life.

Over time, I put plants into the ground around the pond—evergreens and shrubs like japonica, holly, and rhododendron—and set aquatic plants in pots—water lilies, pickerel, and a hardy lotus. I installed a filtration system with an aerating concrete fountain in the likeness of a Hindu deity. Jennifer filled a rock terrace below the pond with flowering bushes, perennials, and an assortment of annual blossoms that attract bees and butterflies in great numbers. Nearby, we planted magnolia, American holly, and native dogwood trees. In this way, slowly over time and plant-by-plant, we transformed the rubber-lined pond into an aquatic garden.

I introduced a pair of goldfish to the pond, where they grew in size until, we suspect, a raccoon killed some of them. The raccoons living in the forest must have seen the possibilities that lie in the pond. Fortunately, the founder fish had reproduced, and several generations of fish now live in the pond; all were born into it and know no other world. For the goldfish in my garden, the pond is the world. They glimmer in gold, red, and speckled-yellow colors as they swim through the water, spinning their fins and flashing their fancy tails. With the fountain running, lotus blooming, goldfish swimming, scent of Magnolia blossoms in the air, and the morning light just so, sitting with George by the pond in the morning with a cup of coffee is like spending a quiet moment on the Indonesian island of Bali.

It is one of the loveliest places on Earth, with steaming volcanoes, lush tropical flora, and rich human culture. When I visited Bali, disembarking the plane was like stepping into a lush terrarium: waterfalls, dripping ferns, bamboo, and flowering trees greeted me everywhere on the island. It's not only nature that makes Bali so lovely; people living there place great importance on keeping their island beautiful. Hugely talented in the arts, music, and gardening, the Balinese take pains to ensure that everything is just so: the delicate ways in which my simple meals were prepared and presented, with fruit and vegetables cut into the shapes of flowers and songbirds; serene architecture of a pagoda temple in the dawn light; musings in a family courtyard; rock walls seemingly alive with sculpted dancers and monkey gods; and the heady scent of incense that always accompanies the performance of a Gamelan orchestra inside a Balinese temple. I imagined how these creative forces were in synchronicity with the tides and trade breezes pulsing across the island, the rushing streams and the palms waving in the wind. It all seemed to be an ageless pursuit of an ecological aesthetic. Art and nature, in fact, have acquired a sacred purpose on Bali. They guide people in the proper conduct of society and extend even to the ways the farmers sculpt their rice paddies into emerald cascades flowing down the volcanic slopes as glistening, fecund examples of terra art. The rice paddies on Bali are so exquisite as to appear on the tourist maps as the "Number One Visitor Attraction."

Water is diverted from Bali's volcanic lakes and shunted down the mountainsides in hand-hewn rock tunnels and earthen canals to irrigate the rice farms below. Majestic pagodas built on the shores of the summit lakes, in the irrigated fields, alongside stream confluences, and finally, at the spot where the spent water empties into the Java Sea, synchronize the sheen of water to seasonal rains, the movement of celestial bodies, religious rituals, and the rice-farming calendar. Where it enters the sacred ponds of the water temples, priests purify water for its life-affirming qualities. From volcano to sea, the Balinese farmers work like mad musicians with their limpid melodies, borrowing from an ancient leviathan score to make the land productive, while the priests passionately conduct their exotic rituals in their water temples with all the flair of symphonic conductors. It is a grand performance. Laboring together in pagodas and rice paddies, lock in step with the hydrological cycle, the farmers and priests are to water on Bali what the Japanese are to rocks in a garden and the Swiss are to watches and clocks. In their masterful hands, water becomes a blessed thing, the holiest substance on Earth.

The water temples wrap religion around water such that it lubricates the cosmological gear works. Water seeps into the flourishing eddies of Balinese culture and land. Farmers sprinkle holy water on their fields to ensure a bountiful harvest. Each day I encountered people walking on the village lanes on their way to a water ceremony, striding along barefoot to a water temple, dressed in fetching garb—golden head-bands and silken waistcloths—accompanied by drummers, flutists, and the fragrance of incense, balancing atop their heads great silver platters heaped with flowers and carved fruit, carrying copper vessels filled with water. In these ways, Balinese society oozes water, as does its natural landscape. The drenching rains that come with each monsoon fill the caldera lakes that nestle within the volcanoes. The encircling sea pounds against the shoreline. Irrigation canals flow down volcanoes and through rice farms, transfusing the land and fostering consanguinity between the island's natural and social histories. Water, it seems, is everything on Bali.

Of course, water is essential for my goldfish pond in Wolf Gap Holler. The pond's natural history began when I first filled it with water. I started its ecological clock that day just as surely as if I had wound its lymphatic spring, set the timer, and called the frogs in for a drink. Such an abrupt prelude is an aberration of nature, where incremental change is the norm of an ecosystem, excepting, perhaps, when a large disturbance occurs, such as a great flood, fire, or winter storm, which then signals a new inchoation, and so, from its outset, my pond was distinguished from the surrounding forest by a sense of time. Its natural history is more closely tied to my own personal history than to any evolutionary trajectory. This gives it a magical quality, not in the way of mythological origins . Rather, I built it, and not because I perform sacred rituals at the pond like the Balinese do in their gardens, but because the timeline of my goldfish pond is melded into my own life at home. I conceived the pond and maintain it, and this makes me its steward. My pond and I now have intersecting lives; we've come to inhabit the same place and time. I am responsible for it.

If I acknowledge my own life to be sacred, how can I not consider the lives and places that entangle with mine to be the same? Such a divine calculation of life lies behind the Deep Ecology School of environmental philosophy, which proposes humankind to be a part of the natural order. It embodies the Arcadian ideal introduced by the eighteenth-century naturalist pastor, Gilbert White, when he described how spiritual forces flowed through his rustic Christian parish in England. It underlies the sentiments of early romanticists like Henry David Thoreau, as well as the later transcendentalists who upheld a natural mysticism in human consciousness. It frames the world views of Buddhism-inspired Tibet and Hindu India. I know my pond is a thing I built with my hands. It didn't just appear out of the blue. Granted, the pond began with a winter storm, which might be considered an Act of God, but otherwise it arose from my imagination and a certain amount of hard work in the garden. It's become naturalized over time, though, so that it functions in accordance with the laws of nature. A premise of Deep Ecology is that sanctity is to be found in nature. Mine is not a wild pond—I take care of it, but it is ecologically alive and spirit-nourishing, and that, I think, is pretty magical indeed.

Some of the finest water features I encountered on Bali were in its gardens. Lotus-laden goldfish ponds filled the landscape and shimmered in the tropical light. Temple gardens contained sacred ponds set into a cosmological design. I sat in family courtyards beside a concrete basin that sufficed for a lake or even an entire demon-filled ocean. A Ficus tree in a clay pot represented a forest grove replete with prayer flags and ancestral spirit shrines. I entered Royal Balinese gardens that always held a goldfish pond as its centerpiece. Even the luxury tourist hotels located on the coastlines boast extravagant water gardens and also some of the most ornate horticultural designs in the world. The Balinese believe that life-giving water contains the strength of the feminine spirit and thus offsets male power attributed to the towering and destructive volcanoes. Maintaining water in a garden is crucial to ensuring harmony in the world by balancing these complementary forces.

In safeguarding nature, art, and human spirit, the Balinese maintain beauty on their island. They believe they have no choice in the matter. It is a person's solemn duty to hold the world together, and this can only be accomplished by striking the right balance in thought and deed. Such a balance, which the Balinese call *alango*, can only occur when the world is kept beautiful. This remarkable idea builds upon the notion that a beautiful thing is as good as it possibly can be; that truth, harmony, and perfection are all to be found in beauty; that beauty, indeed, is what keeps a person in spiritual poise and the world from falling apart. It is this fundamental notion of beauty and balance that inspires Bali's artful way of living. It engages the natural elements—sky, volcanoes, and oceans—alongside aesthetic pursuits—painting, music and dance, sculpture, even rice farming—and finally comes to rest in the garden, where reverence and devotion gather around a goldfish pond as each new day commences.

A Balinese garden may strive in its perfection to emulate the world in balance, but *alango* is not a static posture. In the Balinese gardens, frangipani, gardenia, and night-blooming jasmine suffuse the air with exotic fragrances, but soon their blooms wilt and are replaced by others. Lotus and lily blossoms splash colors across the surface of a goldfish pond with all the existential power of a Gauguin painting. Koi and fancy goldfish stir the water, causing rippled hues that sparkle in the garden. Bali is known to its inhabitants as the Dawn of the World, the place on Earth where time is thought to begin. Accordingly, at first light the Balinese enter their gardens with incense in hand, pray by the pond, and pick flowers for nearby shrines. In Bali, the harmony of the world may well hang on the beauty of a goldfish pond in a garden.

Not only Balinese revere water as a holy substance; many world religions consider water to be a source of purity. My Catholic upbringing in the American Midwest included Sunday Mass, when I'd enter the church and dip my fingers into a bowl of holy water to mark my passage into a sanctified realm. Later in the ceremony, the priest would make his way through the congregation of kneeling parishioners, sprinkling them with water in an age-old blessing known as aspersion. Near Wolf Gap Holler, local evangelicals use a stretch of Red Lick Creek for baptisms by full-body immersions in the water. I sometimes encounter them when driving to work in the morning, greeted by the sun rising over the Kentucky knobs, jubilant cries of "Hallelujah" and "Praise the Lord" coming from the river. I see the mystery of faith time and again in these Red Lick baptisms, which are conducted downstream of chemical-laden farms, straight pipes, and the discharge of household effluents. The Red Lick Christians find a ritual purity in the polluted Red Lick Creek, just as the Varanasi Hindus do in the brown currents of the Ganges River. Elsewhere in the world, too, water has been used for millennia for cultural purposes. In the Biblical lands of the Middle East water distinguishes sacred and profane places, surely a troublesome distinction when it comes to the modern-day boundaries of Israel and Palestine. Jews once used holy water to test the purity of a woman accused of infidelity. Muslims ritually bathe in a splash of water prior to entering a mosque. Water was used in Eastern Europe as a weapon against vampires.

It seems the Hindus have exalted water to its highest post, and sacred rivers and lakes abound in India. I once visited the western desert of Rajasthan to document a religious festival that occurs each year on the autumnal full moon when hundreds of thousands of pilgrims descend upon the tiny town of Pushkar. Over the course of four days, the penitents conduct rites and bathe in a holy lake. One morning I awoke early and hiked to the far side of the lake to make some morning photographs of the temples that ring the water. An outline of hills framed its turquoise surface amid a great offing of magenta sky. The lake began to subtly glow from within as the dawn light intensified. As I set up my camera, I was approached by security officers who ordered me to immediately pack up and leave the premises. They told me I couldn't take a picture of the lake. I understood it wasn't right to photograph the pilgrims bathing at the temples — signboards at such places informed people that picture-taking was prohibited — but I was on the opposite side of the lake, far from anyone, and photographing only the water's surface, not its temples or bathers. I left as requested, but couldn't make sense of the order.

Later that day I went to a local bazaar to purchase some music. A crush of pilgrims filled the streets, making it difficult to navigate through the busy marketplace. Loud music emanating from a bank of monstrous-sized speakers led me to a store where I found a pony-tailed man lounging behind a small counter, his sandaled feet propped up on a table top, fiddling with some knobs of a CD player. I asked him about some gypsy music. The man nodded, pulled a few cases from a shelf located on the back wall, and put a CD into his system. We talked as the music blasted from the speakers. I inquired about the festival taking place outside his shop, asked whether or not the crowds were good for his business. He replied it didn't matter, it had been going on for centuries, and that commerce was a part of it. I made other inquiries. Politely responding to my questions, the man seemed to know a lot about life in that tiny religious town. I eventually discovered that when he was not in his shop selling music, he could be found at the Varha Ghat—Pushkar's holiest bathing site, where he served as the temple's high priest. Each day during the festival, the man conducted a sunset fire worship, called Maha Aarti, on the lake shore. He explained that the male descendants in his family had been performing the rituals for the past six hundred years, and as eldest son it was now his turn to orchestrate the ceremony.

Normally, Pushkar is a sleepy little place and attracts only a minor drift of foreigners drawn to its spiritual vibes. During the festival, though, the town's population explodes to half a million persons, and the surrounding countryside is filled with camels, gypsies, pilgrims, noise, and blowing sand. The priest cum shopkeeper asked what I was doing in Pushkar at that busy time. I explained to the man that I was there to make photographs, outlining my intentions as best I could, and related to him how earlier that morning I'd encountered the police on the far side of the lake. The man listened to my account, smiled, and explained carefully, so that there would be no misunderstanding, that during the four days of the full moon festival, the lake was in fact a radiance of the divine, that during the rest of the year it was simply *maya*—the Hindu concept of illusion—and if I were to photograph Pushkar Lake on one of the festival days, I would, in fact, be taking a picture of God.

That evening I made my way to the Varha Ghat to attend the Maha Aarti and observed the CD seller in his sunset role as temple priest. The lakeside was filling with pilgrims. Peasant women in shawls sat shyly on the marble steps beside flirty gypsy girls wearing frilly skirts, makeup, and glittering costume jewelry. Buxom matrons flashed pearls and wore expensive silk saris. The women sitting together on the ghat looked like variant species of exotic butterflies. Opposite was seated their male kinfolk: a bank of solemn-looking men with sun-blackened faces and flashing white teeth. The men all wore the same style of long-sleeved white shirts, buttoned tunics, and checkered lungi, with tightly-wound bundles of scarlet cotton piled high atop their heads in the distinctive turbans of the region. Identical handlebar mustaches gave them the appearance of having all gone to the same barber earlier in the day.

The atmosphere was both jubilant and reverent as people filed onto the ghat. A trio of musicians appeared in a temple turret above the crowd and began to play an ancient raga on their harmoniums and drums. The devotional refrain soared above the assembly and wafted in a light breeze across the lake toward the saw-toothed hills in the distance. The sun quickly set behind the ridge, and the desert air chilled. Blue and red lights appeared among the temples that lined the shore. Some people threw flowers into the lake. Others lit candles and set them afloat on tiny paper wafers. The surface of the water mirrored the violet sky, intensifying in color until it became a dark pool sparkling here and there with pinpoints of light reflected from the waterside shrines and floating candles.

The white-robed priest stood behind a makeshift altar and led the crowd in chanting. Hands clapped in time to the sonorous music. He had let his hair loose, so it flowed down his back, and when he lifted his arms toward the sunset in an exalted pose, the man's silhouette was as timeless in that old desert place as a stone cenotaph built atop a sand dune. An attendant handed the priest a massive copper disc of burning candles. The musicians played faster and louder, accompanied now by a more spirited hand-clapping and singing, as the priest held aloft the fire disc, spinning it on the fulcrum of his body, offering it to the four cardinal directions before taking aim at the western horizon. The music abruptly stopped, as if Earth had frozen on its axis, and the priest let the candles fall one-by-one into the lake, their flames extinguished amid tiny, hissing spouts of steam. The waxy stubs of the spent candles floated amid the flower petals and diluent deities. The pilgrims filed toward the priest and took their turns to obtain his blessing.

When George and I sit by my goldfish pond in Wolf Gap Holler, we know it is not divine in the way of a Pushkar Lake, nor worthy of the propitiations of a water temple in Bali. We view its components clearly—the water, rocks, plants, fish, frogs, and sunlight—finding a certain pleasure in sitting among them, an old-fashioned entertainment. In its ecological arrangements, my goldfish pond extends far beyond my initial plans for it. Somewhere along the line, it spun out of my design and into the set of rules that govern life on Earth. Even so, the pond's natural history remains tied to my own personal history in Wolf Gap, and thus, in part, is managed by my own memory of it. George and I are not bewildered by that. We get confused, though, when we scale things up, move beyond the pond and consider how our lives and those of humankind in general are in keeping with the state of the planet. When we climb that philosophical ladder it leads to a question: when on Earth did human consciousness and natural history go their separate ways? I have only to look around me to see the answer: never. But we have let civilization get in the way.

For most of human history, people lived in the wilderness as hunters and foragers. Keeping nature intact was tantamount to survival. We needed wilderness because we still were a wild species. Sometime around twelve thousand years ago, we learned how to domesticate plants and breed animals, control water for irrigation, and amend the soil in order to make it more productive. We became farmers. We discovered how to better manage nature to make our lives more secure and comfortable. A few hundred years ago, most of us left the farms for cities and became factory workers, concerned no longer with merely tinkering with nature in order make it more productive but with bending it to our will, controlling it with our technologies, and making it pay. We no longer were content with just meeting our needs as we became fixated on getting what we want. Along the way, we've lost sight of an important part of ourselves: the natural part. As a species, we know by birth what we need, but how are we to know when our desires have been achieved? We have come to identify ourselves with what we possess, and there seems to be no limit to it.

Humankind has the ability to not only control nature but to utterly transform it to meet our demands. This power has made us arrogant. We've learned how to change the chemistry of the atmosphere and the energy balance of the planet. We occupy most inhabitable spaces on Earth, arranging them to accommodate our desires so that very little remains of intact wilderness. We alter the planet's water, nutrient, and carbon cycles in the course of an ordinary day at magnitudes unthinkable only a generation ago. We hold the world hostage to our demands as if we were a gang of thugs, when truly it is our only home. This power has also made us vulnerable. It has made us believe that we are in control of the world. Blinded by this hubris, we've forgotten what we need to know about beauty and balance and can't remember who we are as a biological species. We live thoughtlessly and in ways that give rise to a world that is insalubrious to many people. We no longer rely upon our instincts. Our collective lives may no longer be sustainable. There is no one telling us that we already have all we need. We're instructed to want more, that this will make us happy and keep our societies afloat. This all leads to what may be the most pressing concern of our time: we've become sadly disoriented. Evermore disconnected from one another, distracted, and without due regard for the natural world, or even for ourselves, we run the risk of squandering the gift of the planet that we are beholden to bestow upon our children. What on Earth will future generations think of our stewardship?

If someone from a future generation actually were to stumble upon my goldfish pond, might they not be forced to scratch their head and think: what is that doing here? I've designed my garden pond in Wolf Gap Holler in part to evoke a tropical landscape. It is odd, of course, because I live in a temperate climate. It might have something to do with my desire to escape Kentucky and travel to the lush world of Bali at a moment's notice. It may be just an eccentric preoccupation of mine to mold a tiny part of my world into a place that exists only in my imagination. Or it may have to do with the changing temperatures on the planet. Am I just fossicking in the garden? I know I am pushing the climate map a tad northward with some of its plantings. I sometimes wonder if I'm not, in fact, surfing a global climate wave when I place delicate plants in protected niches in my garden and expect them to live at such a remote distance from their places of native origin. It's a risky business. Some of the plants will do fine, but others won't survive at all. A sustainable ecosystem, like a life, requires a conformation to limits. When I positioned a rubber liner in a hole in the ground, filled it with water and introduced a pair of goldfish, and then cultivated the land around it with non-native plants, how natural is that? Just how natural is the natural history of my goldfish pond?

I visited a lot of gardens in Bali and took notes and made photographs of some of them. I refer to these records at times when I'm working around my goldfish pond, and as a result, its surroundings have become an unlikely palette for plant specimens that evoke an equatorial place on the far side of the world. I can't successfully grow frangipani or gardenia where I live in Kentucky—you can only stretch a climate zone map so far—but I've planted magnolias, which are a good substitute, since they, too, are lovely trees with glossy green leaves and fragrant flowers. The hardy lilies and temperate lotus in my pond evoke the equatorial ones I saw in the Balinese garden ponds. I put goldfish in my pond instead of Koi, which populate the Asian gardens, because Koi tend to root among the water plants and make a mess of things. Jennifer's flowers, meanwhile, attract myriad winged creatures that add tropical confetti to the air when they flap around in it. I put a small glass fishnet float that I found on a beach in Hawaii into the pond. It most likely drifted from Japan across the Pacific Ocean to reach Hawaii, but it's stuck now in my goldfish pond in Kentucky, where it catches sunshine and turns it into an Appalachian rainbow.

Apart from the fountain and glass float, my goldfish pond is composed mainly of natural things, but its history is almost entirely one of human intervention. I'm not even sure it's correct to use the term natural history to describe it. It's more like cultural history I'm observing when I sit beside my pond in the garden. Its botanical components involve colonial plant expeditions, trans-oceanic trade and commerce, greenhouse propagation, selective breeding, and the prosaic matters of soil amendment, weeding, watering, and pruning. The appearance of the garden, finally, has little to do with native habitat, survival of the fittest, or even ecological succession. I try to replicate some of the natural parameters of a habitat, but, alas, in the end, the plants are in the garden because I put them there. Water is in my goldfish pond because I put it there. And the goldfish came from a gardening store. Only the frogs, it seems, are the fully natural inhabitants of my pond. ❖

– Lesson One: How to Build a Goldfish Pond –

The first task when making a goldfish pond is to scout a proper spot for it in the garden. The scale of a pond should fit the space we have for it, just as our lives should match the place where we live; the two become inseparable. In my case, a winter storm created the initial hole in the ground, and I simply shaped it to the dimensions I desired. We could buy a plastic, pre-formed pond in a home supply store and set it somewhere in the garden without giving much thought to the matters of fit or scale, or even to nestling the pond properly into earth; we would pile dirt and rock around it, thinking that such a mound will somehow make it seem like the pond is in the ground. If we simply don't want to dig a hole, or maybe can't dig a hole because of rocky earth, local zoning laws, or an aching back, then I suggest going with a small freestanding container pond rather than a more elaborate liner pond designed to be set into earth. The container ponds look great on patios and fit just about anywhere inside a garden.

If we want a sizeable pond of our own design, though, we have no choice but to dig a proper hole for it and line the cavity with a non-porous fabric liner that holds water. A pond's basin determines its future ecological potential. In shaping it, we might mimic the surrounding organic forms; for example, the habit of a nearby shade tree, undulations in the terrain, or the suggestion of a passing cloud. Imagine the future pond as being alive, snuggling into the skin of the planet and reflecting the sky like a limpid window onto the beautiful world that lies beyond the garden. A goldfish pond can be one of nature's many manifestations, and it, too, requires a proper setting in order to flourish. When building a pond, we create a world, and we need to take proper responsibility for doing so; putting a goldfish pond in the garden is not the same thing as, say, placing a swing set or a picnic table in it.

Once a hole has been dug, we need to buffer it with padding material to protect the liner from possible tears from sharp rocks. This is important even if it seems after digging the hole that there are no sharp rocks near the pond. Don't be misled by their absence. Apart from a sand dune, most garden soil contains rocks, and they eventually will emerge from the glebe and creep toward your pond. It's not exactly tectonic uplift at work, but rather the slow and inexorable drift of dirt under the combined forces of erosion, seepage, and gravity. If we tried to watch the rocks move in the ground, we couldn't see it; just like watching a tree develop—it never seems to grow, yet it does. In any event, we don't want pieces of Earth's fractured lithosphere muscling into the pond.

When I built my goldfish pond, I used pieces of old carpet to line the hole before setting the liner and filling it with water. This raises a red flag: carpets surely contain all kinds of foreign material that can't be good for the natural environment. Still, just about every pond owner I've discussed this with admitted to using carpet at some point as a protective sub-liner for their pond because it works so well. I simply took their advice. The stuff seemingly lasts forever, owing to the fact that it's made from components that decompose over geological time, and carpets are malleable so they can be easily maneuvered to fit the contours of a pond wall. Thick carpets do a superb job of padding the earthen banks, keeping at bay any sharp rocks that escaped detection or that might travel in the ground to the pond despite all best efforts.

It might help to imagine we are making a bed in the ground for the pond, and like a kid living at home under the watchful eyes of a parent, or a soldier serving in the Army under strict orders, we need to tuck in all the corners of the carpet, pull it taut so that the wrinkles and undulations are limited, and keep it free of fallen debris. Don't worry, though, about bouncing a quarter on it like a soldier might need to do in barracks. Most people won't throw coins into your goldfish pond. It's not a wishing well, though it may indeed become a place of good wishes. Silver dimes lying on the bottom of a gold pond would glint, and I find that sometimes I lose my train of thought when I see something bright and shiny. This runs counter to the purpose of a goldfish pond, which is meant to be a place to contemplate the world.

Once we've properly padded our hole in the ground, a liner is set into it. A good rubber liner can be costly, but it's best not to skimp on this because the liner is what holds the water inside the pond. A leaky pond does no one any good. The liner should be 45 mil (0.045 inches) thick, and big enough to more than cover the entire cavity, thus compensating for miscalculations, which are easy to make given the surface we're covering is three-dimensional. Trigonometry isn't needed for this, but we need to carefully measure the surface area of the pond plus its depth and allow an extra third of material to account for sloping walls, plant ledges, and faulty arithmetic. Stretch the liner as tight as possible, and let it overhang the edges of the pond by at least several feet. The overhang is necessary because once we get it properly placed and pulled taut, so that we don't think it will move around, the weight of water will pull the liner toward the center of the pond. Spread the material smoothly with hands and bare feet, moving slowly and evenly like a spider across the bottom of the hole from the center outward. Then let the weight of the water do the rest. It will look fine. Don't worry about a few creases because they'll disappear once we've placed stones, plants, and water in the pond and the liner is covered by a thin deposit of sediment.

The next stage in goldfish pond construction is the placement of rocks. All garden ponds benefit from rocks. While it's hard to find a truly ugly one, it takes a certain eye to see a handsome rock. Some people are born with an affinity for them, knowing rocks to be elegant things, while others can't seem to get past the notion that they are just rocks. Most of us, though, will eventually recognize the beauty inherent in rocks once we spend sufficient time with them. In any event, we need rocks to anchor the pond liner and to create a barrier to the drift of soil moving through the garden. Rocks thus support a pond's infrastructure while also protecting it from an encroaching world of dirt and debris. They also ground the pond into the earth by visually sinking it into a place; if we've done our job properly, the goldfish pond will look like it's been in the garden forever.

The best rocks are those having highly textured but not flamboyant shapes and a colorful patina of lichen, moss, and fungi. These things come from age and prolonged exposure to the elements. A good-looking rock has the wizened face of a pilgrim who's been long in the sun or a Tibetan nomad accustomed to life in the wind and cold. Such faces suggest worldly experience and hint at wisdom and stillness. Highly prized rocks of these shapes are a hallmark of classical Chinese and Japanese gardens, which offer some of the finest contemplative designs in the world, and Japanese gardeners are among the best rock handlers in the world when it comes to placing them inside a garden. The Japanese give names to rocks depending upon size and shape, and position them in a landscape with the precision of a watchmaker inserting tiny gears and levers into a delicate timepiece. The Japanese are to rocks in a garden what the Swiss are to dials in a clock.

There's no need to turn a rock quest into a religious affair, however, as the monks do in Zen gardens in Japan or the Neolithic tribes might have done at Stonehenge or at other ancient sacred places around the world. Still, we should know that in selecting rocks to grace the goldfish pond, we are harnessing the power of gravity and carefully realigning fragments of the Earth's crust. Thus, we create in the garden an inexplicable sense of gravitas. The experience of building a goldfish pond will vary according to where a person lives, and those residing in large cities might have difficulty obtaining the right kinds of rocks. Cities are made of concrete and glass, which obviously have no natural place inside a garden pond, and so a city person might have to buy rocks at a quarry or landscape yard. Otherwise, whenever possible, take a drive in the countryside and scout for rocks in the landscape. I find it to be a great pastime. One might at first feel like a thief taking rocks, but that's only because we imagine others, like ourselves, value a good one when in fact few people even notice they exist. We shouldn't, of course, extract rocks from private property without first asking the owner's permission or from public lands with signboards prohibiting their removal; this goes without saying. Depending upon where one lives in the world, then, garden rocks will be easy or difficult to locate, appropriate or not for a goldfish pond, and derived from different geological histories so that they vary considerably in texture and color. In any event, once we see a rock as a thing of beauty, we can never go back to the old idea of "it's just a rock."

Good landscape rocks have scabrous but not sharp edges and take forms that settle easily into the terrain. They need to be sufficiently large to have a presence in the garden — rocks near a goldfish pond want to be noticed. We should dig a hole for each rock that is deep enough so that up to a third of its mass will be underground. Some people may think that's a waste of hard-earned rock, but in fact a rock that appears to sink into the mantel of Earth, so that only its topmast shows, suggests something that is bigger than it actually is; in this case, a protrusion of the planet's crust. A well-placed rock in a garden is like an iceberg in an ocean. Its mystery is compounded by what is left submerged and unseen. Beyond those fundamental criteria, the selection of a particular rock is, like so much in life, a matter of personal taste.

I would note that, although tempting, the angular rocks found on roadsides where engineers have cleared right-of-ways are not always the best rocks for a garden. They are easy to retrieve, but because they have been fractured by human effort and are devoid of erosional character, they look new and human-made. Even those rocks will do in a pinch, though, and one can always speed up the aging process by coating them with a mixture of yogurt, beer, and moss or lichen gathered in the nearby woods. This is an old trick: put the ingredients into a blender in about equal measure, frappe, and then paint the mixture onto your rocks with a brush. The concoction will fast-forward the patina stage of a rock and forgive the common error of putting a newly blown-up boulder in a pond just because it was a convenient thing to do.

It's important to arrange rocks around the goldfish pond so that they appear to be there naturally. Stagger the plane and linearity of their placement. Put a few specimens of different sizes in a cluster as they might appear on a knoll or along the banks of a lake; Japanese gardeners work in groups of three or five to achieve a naturalistic effect. It's best to stick with rocks having the same general geological appearance, rather than mixing up a lot of different textures and hues: when we finish with our rock garden it shouldn't look like a fruit salad. Allow room to tuck in a few plants between the rocks and arrange things to disguise the edge of the goldfish pond so that its liner doesn't show. Retain some smooth rocks to place inside the pond, but don't worry about covering the entire bottom with rocks, like I did, which is unnecessary because the liner won't show once sufficient sediment settles onto it. A few rocks placed here and there on the bottom, though, will add dimensionality to the pond, and including a few tall ones that stick out of the water will give it a natural look. The frogs in your pond will use the exposed platforms to sunbathe and as perches when hunting insects.

After digging and shaping the depression, installing the liner, and placing the rocks, it's time to add water to the hole. This last ingredient is what makes a hole in the ground a pond. It won't take a pond very long to fill — unless it is preternaturally large — and as the water level rises, we can watch it take shape and come alive. The surface soon mirrors the sky in a bowl of translucent blue light, sunshine, and reflected clouds, and the bottom appears to plunge deep into the Earth's crust. When these events unfold we understand that all the hard work has been worth it. It sounds hyperbolic, given the fact we are filling a hole with water, but the moment is a transcendent one. Enjoy it. A person will be putting the finishing touches on their goldfish pond for the rest of life at home. It will become an essential aspect of the garden. One can mark the seasons of a year by it. It will attract butterflies, turtles, and frogs. We can sit by it and drink coffee in the morning, be entertained, and briefly escape the commonplace distractions of the modern world. The goldfish pond will become a calm oasis in the garden. We might even find a world within it. ❖

CHAPTER FOUR

Garden sanctuaries can be found in the most unlikely places, including in war zones and on rooftops overlooking cities swarming with millions of restive people. Always, though, it becomes apparent in coming upon one that it is there for a specific reason; garden sanctuaries are not accidental places. They are meant to be reflective places where a person can feel safe in the world.

Sanctuary

The gardens around my goldfish pond contain different arrangements of space and light. Walkways and brick patios link one garden room to another, and each to the cedar-and-stone buildings that compose the architectures of our home. A shaded area enclosed by a bamboo grove, shale walls, and cedar fence made from trees felled by a snowstorm hold a stone pathway lined in shaggy mondo grass. Along the way are Oriental pine trees pruned to a cloud shape, rocks, ferns and moss, prayer tablets from Tibet, and small, unobtrusive statuary I obtained in Bali, Nepal, and Thailand. I purchased a large figurine—a fat-bellied, smiling Chinese Buddha—at a nearby roadside stand called Concrete Critters. The statue sits in a corner of the shade garden beneath the overhanging branches of a maple tree. I call this garden my Oriental Garden because the primary influences on it come from what I've seen and learned in the classical gardens of China, Japan, and Bali. Some visitors refer to it as a meditation space, but I think a person can do anything they want in there; I don't subscribe to the theory that a garden has an exclusive purpose.

We have a sunny area that Jennifer tends as a lovely English cottage garden. It commends the value of carefully mixing flowering plants in such a way as to appear random, as they might grow in nature. Her flowers are beautiful beyond belief. The glorious profusion of Jennifer's garden stands beside the starker discipline of my Eastern inspired garden. I have no proof that our respective horticultural realms reflect contrary spheres of our minds, but Jennifer occasionally tucks gay little flowers into the edges of my garden, thinking I won't take notice. From a traditionalist's point of view, bright colors in an Oriental garden are distracting and unsettling, to be used sparingly and only then to mark the seasons and passage of time. An English Garden, on the other hand, is meant to celebrate the profuse tinctures of life. Not to push this analogy too far, but my Oriental garden imposes order on the world while Jennifer's cottage garden embraces the exuberant variegations in it. Beyond these spousal-differentiated gardens, buttressed by a stone terrace wall, is a vegetable and fruit patch. Some of our home gardens are loosely-designated "her" or "his" spaces, tended by either Jennifer or me, but usually not by both of us working at the same time — unless, perhaps, some heavy lifting is required. We both like our space and the meditative silence that can be found in a garden and wish to avoid gender collisions while in them.

On the far side, where a pine grove once stood, is my goldfish pond, and beyond our cultivated gardens is native forest. It, too, lightly bears the mark of our hands; we pull the invasive honeysuckle that strangles the young saplings and move small trees around to fill openings in the canopy. I like to think of a garden as a kind of installation art. I impose my design and imagination on them, but they are free to follow the genetic predilections of their botanical elements. In the end, cultural aesthetics pay homage to ecological reality inside a garden.

Many people see gardens as sanctuaries. The word suggests a place of refuge. Spiritual traditions around the world embrace sanctuaries: altars where the presence of the divine is felt, sepulchers containing the relics of saints, or natural areas set aside for worship, reflection, or prayer. This is not to imply that gardening is a kind of religion, although for some people it is. A sanctuary in the Christian tradition is marked by symbolic architectures such as altar rails, naves, and iconostases. These distinguish it from secular spaces. Gardens, meanwhile, have fences that might serve a similar purpose. In Tibetan monasteries, the sanctuary is an inner sanctum where statuary and the relics of bodhisattvas are safely kept. These interior chapels, covered in tantric paintings, open onto special entryways that may be accessible only to high-ranking lamas. Most gardens have gates of one kind or another. The inner sanctum of a Tibetan monastery, marked by the intonations of chanting monks, gongs, and ringing bells, is also impregnated with the redolence of aromatic incense and sour-smelling butter lamps. Gardens, on the other hand, are filled with songbirds and the fragrance of flowers.

The courtyard sanctuary of an Islamic mosque is called a *masjid* and is where the congregation assembles to pray on holy days. It typically is a large and empty room, offering only floor space for prayer rugs under a vaulted firmament punctuated by the spindleshanks of surrounding minarets. Most gardens have benches and open onto the sky. A masjid may be of enormous size. The Masjid al Haram mosque in Saudi Arabia, for instance, covers almost ninety acres and can accommodate a million persons during the Hajj Mecca pilgrimage. Thirty-five acres in the heart of Jerusalem is taken up by the Al-harif al-sharam mosque, also known as The Noble Sanctuary. A garden, meanwhile, may be as small as a single potted plant in the corner of a room. In Hindu temples, priests tend the deity statues. Gardens commonly have concrete gnomes or other whimsical figurines watching over them. In any event, to see a garden as a sanctuary is akin to viewing the natural world as a sacred place.

A sanctuary can also be a secular haven: refuge, emergency

shelter, hospice, homeless shelter, psychiatric hospital. We have political and humanitarian sanctuaries, refugee camps, and asylums. We have sanctuary movements and safety zones for persons fleeing civil unrest. Some sanctuaries are designed for abused women, orphans, or people otherwise burdened by life's injustices. Therapeutic sanctuaries offer spa treatments, massage, and aromatherapy experiences. Taoist sanctuaries teach people about the movement of *Qi* (life energy). Artist sanctuaries help people learn to creatively express themselves. There exist sanctuaries for endangered plants, for wild animals to live until their natural deaths, and for pets, wolves, horses, birds, and monkeys. Even American baseball has a home plate where base runners are safe from the threat of expulsion.

Off all possible sanctuaries, spiritual or secular, gardens actually have the longest-serving history. Prior to Adam and Eve eating from the Tree of Knowledge and thus committing the Original Sin, the entire world was a garden, the Garden of Eden. In this case, a bite of fruitful wisdom unfortunately shut light on the world's natural sanctuary from which we have so much to learn, and we've struggled ever since to know our true place in the world. Still, many ancient Christian sanctuaries and abbeys retained gardens for prayer and contemplation. India's Moghul gardens, filled with bubbling fountains, night-blooming jasmine, birdsongs, and shaded patios, served as private places for the pleasures of royalty but also to hold Islamic worship ceremonies. The great gardens of England, places like Sissinghurst Castle Garden in Kent and others of equal stature, played a similar role in British society; walled enclosures kept the peasants at bay while clergy in the Church of England catered to aristocratic whims. As we have seen, the Balinese temples contained all manner of sacred gardens and water ponds.

The design of a garden's *sanctum sanctorum* reflects cultural tastes. It might hold the meaning of an entire cosmology, such as we find in Japan's temple gardens, or merely the eccentricities of an individual's mind. Garden sanctuaries can be massive or tiny, unkempt or neat as a pin, orderly and geometric, or a riot of whimsical plantings. Some are filled with flower beds, entire woodlands, or, in the case of Oriental gardens, well-disciplined bonsai trees a foot high and many centuries old. Garden sanctuaries can be found in the most unlikely places, including in war zones and on rooftops overlooking cities swarming with millions of restive people. Always, though, it becomes apparent in coming upon one that it is there for a specific reason; garden sanctuaries are not accidental places. They are meant to be reflective places where a person can feel safe in the world.

George and I have little to fear when we sit in my Wolf Gap garden beside the goldfish pond and drink coffee in the morning. Occasionally, we get a bit annoyed when someone makes a racket in the valley, but that's only because it's usually quiet as a church mouse around my place. We've grown accustomed to the sound of silence. Nor do we fret about being hit by a car, mugged, or run over by a bus. It's not likely we would be pried upon or attacked by wild animals, even though the surrounding forest is full of them. We've never discussed it, but I suppose we figure that we could hold our own against the advances of opossum or raccoon and probably a deer, but not a skunk—that bravado may change, though, with the recent reports of bear and cougar again roaming our region. Nor are we afraid of being late for an impending meeting; we allow ample time for our morning coffee at the goldfish pond. I'm fairly certain that others might also find the aspects of a sanctuary at my goldfish pond—respite, inspiration, beauty, contemplation—but a person would have to seek them out.

Some gardens are designed in such a compelling way, such that not knowing they are sanctuaries is hard even to imagine. I once visited the Chinese city of Suzhou to visit its gardens. They are a wonder of the world. Built in the classical Yangtze Delta style during the Ming (1368–1644 C.E.) and Qing (1644–1911 C.E.) dynasties, and numbering more than 270, the Suzhou gardens are so masterfully done that the city was declared a United Nations World Heritage Site in honor of their multitudinous presence. The Suzhou gardens are enclosed by stone or adobe walls, through which I passed via a moon gate. The round doorway suggested I was to leave the grid works of modern society and enter a circular world of illusion, nature, and horticultural magic. I was being asked in a roundabout way to set down my mental right angles, levels, and rulers at the garden entrance, as I would my boots and gloves when I come in the house after working in my Wolf Gap gardens, and to consider a new way of measuring the world. The Cartesian tools of a modern gardener, useful in our logical world of squared-off assumptions, wouldn't be of much help to someone wishing to make sense out of a Suzhou garden.

A chimera of ponds, rock grottoes, and carefully-pruned conifers greeted me wherever I walked through one of those Chinese moon gates. A Lilliputian diorama of mountains, rivers, lakes, and forest appeared at an elfin scale. Stepping into such a garden was akin to somersaulting into a classical scroll painting—one that featured craggy mountains, wind-blown pine trees, and mist: the kind of picture you might find hanging in the lobby of just about any Oriental restaurant in the world. Imagine leaping inside the scroll art just like that: take in a landscape's measure, turn a corner of a cliffside trail, duck under some pine boughs, and become enveloped within a swirling mist that writhes like a dragon across the canvas painting. Embarked upon such an imaginary journey, going deeper into the mind of the artist, we might gain a sense of what it's like to enter a Suzhou garden: puzzling, transcendent, and a hauntingly beautiful experience.

The idea behind a Suzhou garden is to replicate on a diminutive scale the iconic places of nature that hold therapeutic or spiritual powers in traditional Chinese philosophy: holy mountains rising from the diluvium of mist-filled valleys, isolated hermitages, pilgrimage trails, and ancient wizened forests transfigured by the wind. Seeking the geographical roots of the classical gardens of the Yangtze Delta, I took an overnight train to the Jiuhua Shan Range located 250 miles west of Suzhou. It is one of China's four sacred mountains. There I walked amid the contorted limestone outcroppings making up the Nine Glorious Mountains. I halted at viewpoints visited ages ago by painters, sages, and poets whose art had moved the Suzhou gardeners to create their horticultural masterpieces. Such remote landscapes made their way into the city gardens by passing first through the human imagination. From the mind of a poet to the vision of an artist and into the hands of a skilled gardener, the Suzhou enclaves captured the nature-based sentiments of an ancient civilization.

A famous Tang dynasty poet, Li Bai (701 – 762 A.D.), chronicled the Jihua Shan range in verses that are chiseled onto rocks in some of the Suzhou gardens: "Sailing down the Jiujang River the other day, I saw the Jiuhua peaks in the distance, looking like a river hanging in heaven." The poet Li Bai thought the nine peaks of the range were like nine lotus flowers reaching into the sky. The melodic structures of his old-style gushi poems mimicked the mountain topography and attracted mystics, monks, and warriors to Jiuhua Shan, where they built temples amid its rock outcroppings and established important schools of philosophy and the martial arts. The mountains came to symbolize the celestial axis of the Middle Kingdom, from which the Emperor's reach extended to heaven. No wonder, then, that Jiuhua Shan came to inspire so many of Suzhou's masterpiece gardens.

I discovered in the Jiuhua Shan range the capricious rock formations, bamboo groves, fern glades, and gurgling mountain streams that are so widely and reverently depicted in Oriental garden iconography. I witnessed lone pine trees bent against the wind—leaning away from a cliff in the classic posture depicted in scroll paintings—as if planted there by a Suzhou gardener. The most renowned tree in the region is the Phoenix Pine, also called the Number One Pine in the Land under Heaven. It is huge and magnificent and appears to dangle by its fragile roots in mid-air. Dampened by mist swirling up from the valley floor, I treaded a narrow trail in the face of the cliff toward a fifteen hundred year-old monastery dedicated to the God of Earth, imagining all the while that I might have tumbled into one of those restaurant scroll paintings and gotten caught in a cauldron of fire-breathing serpents, flying monks, Tao te Ching throwing sticks, amulets, magic, and talking trees. I visited places with wondrous-sounding names: Taoyan Waterfall, Shutan Pool, Sea of Clouds Around Lotus Peak, Singing Stream, Snow Mantle over Flat Terrace. Greatly inspired by the idea of such places, the gardeners long ago in Suzhou created majestic retreats that continue to hold a mystery as the spiritual heirlooms of a deeply contemplative way of seeing the world.

When the Suzhou gardens were first built, few people had the opportunity or means to visit Jiuhua Shan or the other holy places located in distant regions of China. The gardeners sought to symbolically bring those sites into the city where they could be enjoyed by townsfolk. The gardens were expensive undertakings, though, and commissioned mainly by wealthy merchants or high-ranking government officials. The urban elite desired status and vied with another to create evermore grand gardens filled with metaphysical effigies so finely wrought as to be the best of their kind in the world. The architecture of the natural word became the foundations of the Suzhou gardens: artificial streams flowed through as if emanating from mountain springs; man-made limestone grottoes conjured sacred mountains; the placement of a single culm of bamboo intimated an entire grove cascading down a hillside; a well-pruned tree became a woodland;

stepping stones in a gravel bed were like islands cast into an ocean. Drawing inspiration from the poets and painters who had visited the sacred sites, the gardens were twice-removed from their geographical origins. This opened up room for innovation and creativity. Observed from numerous vantage points, a person inside a Suzhou garden might well imagine themselves to be pilgrims on a long journey to one of Earth's holy places.

I strolled among gardens with beguiling names: Fishing-net Master's Garden with a Pavilion for the Arrival of the Moon and Wind, designed as space within which to contemplate the night sky; and the Humble Administrator's Garden with features such as the Fog Awaiting Pavilion, the Drifting Fragrance Hall, the Mountain View Tower, the Pavilion of Lotus-scented Breeze, and the Chamber for Thirty-Six Mandarin Duck Pairs. Moving among these fetchingly named courtyards, I experienced a sort of shape-shifting cosmology and often felt caught between worlds. I finally suspended any pretense of garden knowledge, exploring, instead, the hyper-imagined reality that exists inside a Suzhou garden.

I've attempted to bring some of the attributes of a Chinese garden into the landscape around my goldfish pond in Wolf Gap Holler. I haven't gone so far as to replicate the rock grottoes commonly found in the classical gardens, but the possibility is there for me to do so. One of the most compelling features of the hills surrounding my home is their geological undergirding, which is comprised of the kind of rock that would make a good-looking garden grotto. The Suzhou gardeners referred to rocks as the "bones" of a garden, suggesting the structural role they play in the anatomy of a garden. The Karst limestone around my home derives from coral reefs that inhabited the Silurian Sea more than three hundred million years ago, and is similar in look, origin, and texture to the Karst bedrock that undergirds the Jiuhua Shan Range, the latter having been similarly shaped by erosion acting upon the lingering presence of a Paleozoic ocean. I often walk among the rocky promontories above my Wolf Gap home, poking into their nooks and crannies, and come across small caves, fossils of ancient marine organisms, stone platforms, and, at the point of a cliff where it is fully exposed, natural fern-laden and craggy gardens filled with mountain laurel and rock formations that would be the envy of a Suzhou gardener. It would be a huge task, of course, to bring these rock ledges down from the ridge tops and into my garden. It would be tantamount to moving a mountain one piece at a time. I've been content, therefore, with leaving the cliffs where they lie, to look up at them from my garden, accepting they belong best where they already are found, and feeling blessed to have their beauty all around me. ❖

CHAPTER FIVE

The sun rising in the east strikes first in the garden at the goldfish pond. George and I enjoy these early rays when we sit on the bench and have our morning conversations. They are warm without yet being too hot. With the coming radiance of the sun, the pond springs into life, gearing up to the divine possibilities of a new day.

The Flow of Energy

I was riding shotgun in an old Peugeot sedan in the countryside near Hanoi with my friend Andy and his two Vietnamese colleagues when one of them asked what I did for a living. Their eyes lit up when I replied I was a geographer. It was a curious reaction. When I'm in America and someone asks the question and I say that I'm a geographer, their eyes generally go dim. Sometimes a person might go on about how when they were a kid they liked to hunt for fossils, and then I feel compelled to explain that I am a geographer and not a geologist, and this only begs the question: *What's the difference?* The confusion is understandable. Most Americans never study geography in school, and so it's to be expected that they'll have little idea about it. I've learned not to pursue the topic during the normal run of a conversation as it only leads to a dead end. So, of course I was interested to learn that my Vietnamese friends understood the work of a geographer. And it turns out to be a respected profession in that country, but for an unexpected reason: a geographer in Vietnam decides where to place the tomb of a recently deceased person, which, in turn, may determine the fortunes of a household. A properly situated grave can bring prosperity, while an inauspicious location might drive a family to financial ruin. My Vietnamese friends informed me that

I'd have no problem finding a job in their country. They explained that it all has to do with geomancy.

In ancient Greece, geomancy meant "foresight by earth" and was equated with the animation of life on the planet. Arabs know it as the "science of the sand." European scholars in the Middle Ages used geomancy to determine the alignments of planets and stars in the sky. In Vietnam, the practice of geomancy is equated with *feng shui*, which translates as "wind-water" and concerns the flow of energy, called *qi*, which connects places and sentient beings on Earth to heavenly bodies. Geomancy proposes that certain spots are more auspicious than others because they better synchronize the flow of energy through the physical elements of the planet. According to *feng shui* precepts, if one is aware of such consanguinity, it is possible to design space, a life, or in the case of Vietnam, the placement of a grave to be in better keeping with celestial events. In Western parlance, we may say "go with the flow" to suggest a similar apperception. I wonder, though, how many people understand how provocative this cliché really is? To go with the flow, it would seem, might deserve more attention than it gets as a mere bumper sticker. But just as I tend to leave aside the nuances of an explanation about geography when I am in conversation with my friends, so too, I avoid discussions of geomancy with my professional colleagues, who tend to be skeptical around ideas lacking a scientific basis.

The practice of *feng shui* emphasizes designing our homes and lives to be part of the natural energetics of a place by making arrangements that not only stream the flow of energy but intensify its positive influence in our lives. Energy is an irrefutable force of our world. But because many of us equate *feng shui* with its commercial side—metaphysical baubles and designer colors—we miss the greater possibilities of learning to live more deeply with energy in mind. Surfers in Hawaii understand it, however, as do snowboarders in the Swiss Alps, sailors in the

America's Cup yacht race, kayakers in West Virginia, meditators in Tibet, solar engineers in Nevada, Kung Fu monks in China, atomic physicists in France, and drummers in West Africa. They all channel energy to get a job done. And gardeners...who might know better than them about the flow of energy? A goldfish pond is a splendid example of it.

Many visitors to our home spend time at first wandering around the place. They might observe its structures or horticultural eccentricities, and then, I imagine, try to get their bearings. It's a matter of predilection. Some people hardly take notice while others poke around quite a lot and absorb the details of a place. When curious people get past the surface of our home and its gardens and settle into them, they often comment on how pleasant it is to be there—the goldfish pond, gardens, and all. It's a nice sentiment, flattering in a way, but I'm not sure what to think about it. Perhaps they've glimpsed the spirit of the place or maybe they're just being polite. A few stumble around the nooks and crannies of my home as if they're lost in a distant land. Some make it over to the goldfish pond and take a seat on the bench. Maybe they are simply tired and want a place to rest. There are so many possibilities. Still, a certain number of visitors to our home remark that it has a nice energy about it.

I live on a small bench of land above the confluence of two streams. The property faces south. To the north is the round hump of Wolf Gap Mountain, an outlier of the Cumberland Plateau. By straddling this physiographic lineament, I live in a preferred *feng shui* site with a mountain to the north, water below, and a south-facing aspect. I once lived in Hawaii near a Chinese cemetery that benefited from similar geographical coordinates so that it was a revered site, and I've spent much time among Himalayan monasteries that exhibit these same qualities. One of the most remarkable structures I've encountered along such lines is the Borobudur temple located on the Indonesian island of Java. Like the Chinese cemetery in Hawaii and Tibetan monasteries in the Himalaya, it aligns with a flow of energy at astral scales. Borobudur's spiraling basalt staircase noses into the sky and leads upward past dozens of life-size Buddha statues and hundreds of carved stone scriptural panels evoking saints, sinners, and the various stages of human life. When finally I attained the summit of the monument, after encircling it many times on my way to the top, I looked out upon a steamy sweep of jungle, river, mist, and smoking volcanoes. From this vantage, the temple appeared to be at the center of a brand new world.

Whether or not an auspicious setting exhibits any special qualities depends, of course, upon a person's intuitions about the world. Such a setting may just provide an interesting angle or a scenic location. At the very least, *feng shui* and related notions about planetary energetics require that we suspend the topographic rules that normally determine the way we judge the contours of our world. Once a person apprehends life in a certain way, though, little may appear again as it had before, including how we view life and the world in which we live. I have a quotation tacked to my office door at the university where I teach that reads, "The idea is not so much to see different places but to see places differently." This pretty much gets at the heart of the matter. It is a bit like peering at the world through a kaleidoscope: with each adventurous turn of the dial the scene we apprehend becomes unhinged from earlier apparitions and reconfigures into something quite new and different. The results may be very strange to a casual observer.

A cemetery in Vietnam, for instance, looks nothing like the ones I come across in Kentucky where the tombstones stand tall and erect amid their colorful arrangements of plastic flowers. The burial markers line up in neat little rows, all facing the same direction, appearing on a Red Lick hillside like battalions of dominoes festooned in colorful Hawaiian leis. In

Vietnam, though, I found grave stones of all sizes lying higgledy-piggledy on small mounds of dirt without adornment. We passed a small Taoist graveyard on our driving tour near Hanoi and I asked my friends to stop the car so I might have a closer look at it. The lichen-covered tombstones angled in all directions, with no apparent order to them, seemingly just piles of rock tablets that might have been dumped together in a spot and left to lie where they fell. They tilted this way and that and looked as though they would topple over at a moment's notice. The affect was of random negligence, but it wasn't that way at all. When I asked about it, my Vietnamese friends explained that the stone markers reclined in various directions because the people whose bones were interred under them had died on different days of the year. Space and time had been carefully calculated into the disposition of the grave stones.

Looking beyond the cemetery, I saw a jade green landscape of columnar hills and honeycombs of deep-set grottoes and wet bottomlands. The tombstones were set in geomantic alignment with those distant crustal formations and hinted in their offing at some deeper meaning in the Taoist landscape. I made plans a few days later to explore this beautiful and bizarre countryside by signing onto a tour to the Perfume Pagoda. It was located some distance away in the Huong Tich Mountains. The famous range exhibits the classic features of a Karst landscape, where fluxion and the dissolution of calcareous bedrock leave behind a latticework of caverns, subterranean streams, and steep palisades with sheer rock faces. Only a few places in the world are as famous for its Karst formations as this region of northern Vietnam. One is the provinces of southern China where rocks were quarried for the grottoes assembled inside the Suzhou gardens. Another is Kentucky.

The Karst ridges above my Wolf Gap home are pockmarked with sinkholes and overhanging precipices lined with moss, ferns, delicate columbine flowers, and the fossilized relics of marine organisms. One of the world's largest caverns, Mammoth Cave is located a few hours' drive west of Wolf Hap Holler. The Kentucky caves share a basic geological structure with the Huong Tich Mountains in Vietnam. For Vietnamese people, though, the rocky precipices are believed to be filled with astonishing sacred energy. Some of the holiest temples in Asia were built inside their caverns many thousands of years ago. Meanwhile, more than a few of the Kentucky caves were used by moonshiners to make corn whiskey during prohibition years and later became convenient dumps for all manner of household waste. It always intrigues me to see how diverse societies uphold very different views on similar features of the natural world.

The first stop on my Perfume Pagoda tour was a small Taoist temple perched on a ledge at the base of a steep mountain. Its pagoda roof gave the structure the look of a bird of prey set to take wing. The parking lot was empty when we arrived, the gate into the complex still closed. Our guide went inside to arrange our entry, and we assembled in a cavernous room that served as the temple entryway. The guide began a long-winded discourse on Taoist history, explaining how the site where we stood exemplified the principles of *feng shui.* He chronicled the temple's lineage of famous Tao teachers and meditation masters, and in a show-and-tell manner explained the purposes of the odd furnishings and ritual accoutrements that filled the space around us. It was a lengthy disquisition, and we all waited patiently for him to finish.

The man explained that the term *Tao* means "Way" and that the "Way" interprets the "flow of the universe." Tao is immanent in all aspects of nature, living or not, including within each of us. He reminded us that the founder of Taoism was a philosopher by the name of Lao Tzu, who in the sixth century B.C.E., wrote a famous treatise called *Tao Te Ching*, in which he laid out the principles of the Taoist point of view. The *Tao Te Ching* is well-known in some western circles, and a few members of my group were familiar with it. The guide clearly relished his role as an explainer of arcane ideas, and embellished his talk with vague references to the ancient Tao treatises while we stood in the dimly lit chamber amid burning sticks of joss and ancient talismans. Some in our group looked askance when the guide quoted a line from Lao Tzu: "darkness within darkness, the gate to all mystery," and then suggested that a gloomy temple, perhaps the very one we were in, had provoked the ancient sage's sentiments.

At about this time, a pair of mute dwarfs suddenly appeared in the room and pointed toward a passage at the back. They whistled, clapped their hands, and frolicked crazily amid the temple furnishings. We didn't know what to think. As the tiny duo bounced around the room, thumping one another on the back like old tennis partners and laughing silently while our guide continued with his talk, some members of our group grew impatient with the tomfoolery. Others were embarrassed by their puckish and misshapen appearances. They certainly were an odd-looking couple. It was obvious, though, they belonged in the temple, which in fact was their home. They paid no heed to our discomfited tour guide, whom they mimicked in grotesque body language whenever he reprimanded them for disturbing his speech. All the while, the mute pair beckoned us to follow them out the back of the temple.

I left the others and followed the achondroplasia men through the rear passageway and into a natural cavern of immense size. We passed through more caves and tunnels that led us deeper into the mountain and reached a candle-lit opening filled with stalactites and stalagmites. It was a surreal journey into the heart of the mountain. The two dwarfs stopped to cavort inside this large cavern, still communicating with whistles but now also banging rhythmically with metal spoons on the cave formations, playing them like a glass harmonica, hopping nimbly from one speleothem to another, their faces lit by candlelight so that they glowed like shining visages held up against the theatrical backdrop of shadowy rock balustrades and dimly lit stone towers. The effect was of the dwarfs putting on a kind of *kolom*, or masked drama, a wild hybrid of Renaissance *ballet de cour*, Indonesian puppet theater, and the spirit dances of old Indochina. They ushered me further inside the cave with their hand gestures and huge, radiant smiles. I wasn't entirely sure it was a good thing to keep going, but the little laughing men seemed harmless to me, and they obviously thought I should see something very important.

I followed the pair toward a tiny hole in the wall that leaked daylight. A lone sun ray pierced the crevice and formed a portal illuminating a strange and convoluted world. The dwarfs beckoned me through the passageway and onto a ledge that was cut into the back side of the mountain. We were on a narrow trail that led up a steep cliff face to the summit of the ridge. I was even less certain now that I should continue: time was passing, the path was getting dangerous, and I thought my tour group must be about ready to leave the temple. My little guides urged me on, though, and pulled at my arms to keep me moving forward. *Come on,* they seemed to say. We reached the end of the trail atop a small tableland hovering above the countryside like an antediluvian sky platform. A glistening panorama of knife-edged mountains and wet gorges fell away in all directions. A river meandered through the convoluted terrain. I sat down beside the dwarfs, who had become quiet of a sudden, and gazed onto a scene that was so beautiful it moved me to tears. All along, the two men had simply wanted to share the sight with someone.

Returning down the mountain, I passed again through the dimly lit rooms located behind the temple and noticed things I hadn't seen before: votive lamps, statuary, offerings of incense and money, and prayer mats. Clearly, the cavern was used for worship and meditation. Here, I realized, was the true temple; the pillared pagoda out front was simply an antechamber to greet guests before they crossed the sacred threshold into a natural sanctuary. The latter was built not by men but rather forged by seismic forces and made holy by centuries of Taoist faith. I would have missed all that had I not followed the mute pair to the top of the mountain. When I exited the temple, I found the others waiting for me in the van in the parking lot. My guide asked me where I had disappeared to, scolding me for being late and holding up the tour. I kept quiet, but wanted to reply that I had been to the ends of Earth and back.

Caves like the one in Vietnam have been used for millennia among diverse cultures as prayerful places. The upper Mekong hills of Laos and Thailand are filled with them. So, too, are the Himalaya Mountains and the Plateau of Tibet. The ancient Ellora and Ajanta caves in India host fabulous Buddhist meditation chapels and some of the most treasured collections of ancient religious murals in the world. The sacred Lascaux Grottoes in France are famous for their mysterious and colorful paintings. The American southwest is littered with caves that host enigmatic petroglyphs and pictographs. Evidence obtained from caverns in Kentucky show that they, too, were often used by native people for sacred rituals. The cave in Vietnam, though, was exceptional in my experience. I was lucky to have been guided into it by a pair of laughing gnomes, to whom I owe gratitude for sharing with me a glimpse into their fantastical world.

The placement of a temple inside a cave, indeed the transformation of caves *into* temples, suggests a way of seeing them other than as a mere topographic defile. Some tantric masters in the Himalaya spend many years meditating inside a cave with no outside contact apart from an attendant who brings them food and water. They sit alone in the cave and focus their thoughts on the interior workings of their minds. Grottoes can be places of haunting beauty. Spelunkers in Kentucky often comment in reverential terms about the experiences they've had inside a cave. I normally feel claustrophobic inside a small cave, but I understand how they might also be chambers of reflection for some people, places where a person could feel close to Mother Earth in a womb-like enclosure. Certainly, the large caves I've been in have always filled me with a sense of wonderment. The Taoist masters believed that a cave is a void where the planet's energy is concentrated so that it might enter more easily into a person's consciousness. Perhaps that is what Lao Tzu had in mind when he wrote that caves hold more than darkness; they might also hold clues to some of life's most perplexing mysteries.

The cognation of geography, energy, and the human spirit is a theme that runs through many ancient cultural traditions, determining how people come to understand their place in the world. It makes nature at once radiant and holy. Such an idea, though, is alien in much of modern society, which toggles awkwardly among mythology, intuition, and the need for empirical verification. Even here, though, we find examples of its application. Western practitioners of *T'ai chi chuan* and other forms of moving meditation seek to activate their *qi* energy for its health benefits by adopting poses that mimic the elements of the natural world. Such postures are essential as well for a mastery of the martial arts. Disneyland in Hong Kong repositioned its main gate by twelve degrees in order to be in a better alignment with *feng shui* criteria. Donald Trump hired a *feng shui* master to analyze the site of his Trump Towers in New York City. A McDonald's fast food restaurant in Los Angeles was renovated in consultation with *feng shui* experts. I've designed my home in Wolf Gap Holler according to what I think about the flow of energy, but I rarely mention that to a person visiting my gardens and goldfish pond. I know it sounds odd, and I'd be hard pressed to come up with any solid evidence for it. In any event, a debate about living with energy in mind may never be easily resolved because it spins atop the messy fulcrum of science and spirituality.

There are a few points, though, on which the mystics and secularists might agree. For example, we know that the movement of energy through space and ecosystems is what makes life possible on Earth. We've become accustomed to thinking about energy, though, as a kind of commodity: fossil fuels mainly, and not as a particle or wave. Fossil fuels simply are a means of storing solar energy: a seam of coal, a puddle of oil, or a bubble of natural gas. They are vessels containing the remnants of carbon-based life that was fueled by the sun and trapped in biological layers over many geological epochs. When we unwisely use fossil fuels, we run the risk of depleting these vessels, exhausting the stored reserves of oil, coal, or natural gas. Furthermore, it's often dirty and toxic when we burn these substances in order to release their energy potential. Toxic gases escape that put our planet at greater risk. Renewable energy such as wind and solar radiation derive from the planet's atmosphere and are among the most plentiful and sustainable forms of energy we have available to use. The *feng shui* practitioners understand this as well as physicists. It is the basic idea behind the so-called alternative forms of energy. But they are *alternative* to what?

Earth's climate moves energy around the planet. The planet's atmosphere, in fact, is an enormous energy system. When sunlight passes through the glass panes of my house in Wolf Gap Holler, it heats the interior and is stored in the brick floors and concrete walls of the structure. This helps keep our home warm and cozy at night in winter. The sun shining in the gardens and forest around my home makes life possible in Wolf Gap Holler. Solar energy also moves through the trophic levels of my goldfish pond. A trophic level is an eating platform, a further step away from the sun, part of a food web entwining all life in a flow of energy. Food chains start at a basic trophic level with plants, which capture energy directly from the sun through photosynthesis. When an animal eats a plant, it consumes the stored energy and thus occupies a secondary trophic level. Carnivores eat at the highest trophic levels, furthest away from the sun. A pond, even a hand-made one, is an ecosystem that captures solar energy and moves it through a complex web of feeding relationships. Energy goes up and down the food chain in the manner of who eats whom and what.

I think about it in this way. My home rests in a cradle of south-facing land situated between a mountain (*yang* — male energy) and a valley stream (*yin* — female energy). According to the principles of *feng shui*, this is an auspicious location because it is a balance point in the flow of energy. It also is part of the philosophical premise of Taoism, captured in the popular *Taijitu* ideogram of a circle split into alternating parabolic light and dark halves signifying quiescence and the duality of *yin* and *yang* energy (☯). Balance is embedded in the design of our home and gardens, which also modulate the flow of energy. My Oriental garden, for instance, is based upon the energy principles of ancient China and Japan, while Jennifer's English cottage garden does a splendid job of harnessing the sun to grow flowers and vegetables. Our cat Olive, meanwhile, spends much of her day in the gardens, moving from spot to spot, always seeking the best possible place to sleep, which invariably means following the sun in cool weather and avoiding it when it's hot outside. Olive is an energy-seeker. She also is a killer by instinct. She hunts and eats at a relatively high trophic level, thus capturing energy flowing from the sun through plants and into rodents, ground squirrels, small birds, baby rabbits, and goldfish. Olive is not a vegan. Her teeth chatter in anticipation of a meal whenever she sees a bird. She's a predator and hunter, and one of her favorite places to stalk prey is the garden.

When it comes to the flow of energy, my goldfish pond is the world writ small. It sits in the *yin-yang* lap of Wolf Gap Holler and results from an energy event, a winter storm. The sun rising in the east strikes first in the garden at the goldfish pond. George and I enjoy these early rays when we sit on the bench and have our morning conversations. They are warm without yet being too hot. With the coming radiance of the sun, the pond springs into life, gearing up to the divine possibilities of a new day. Plants at the pond collect solar energy and transform it into leaves, flowers, roots, and stems. The plants *grow*. The higher-level pond organisms consume the sun-catching autotrophs and store their energy. And Olive occasionally eats some of them. If it's mice and moles, it's generally okay, but when she snatches a bird, frog, or fish from my pond, it breaks my heart, and I am torn again between a fondness for my feline friend and knowing that I've let loose a cold-hearted predator in my garden. When Olive has eaten her full, she finds a warm place in the sun to sleep.

Feng shui, with its attendant notions of energy, has its critics, even in China where the practice first took root. The early Jesuits in China thought it superstitious and absurd. Other Christian missionaries were more adamant; thinking it a heathen practice, they built churches with tall steeples to "puncture" the *feng shui* myth. *Feng shui* practitioners were persecuted in China during the Cultural Revolution of the 1960s for being superstitious feudalists, and some of their rituals were banned because they were viewed as threats to the power of the Communist Party. It's absurd, of course, to equate Earth energetics with a political ideology, but in fact, such misguided notions about power and nature are commonplace in the world. Most people, though, simply see *feng shui* practice as a harmless pastime—if they think about it at all—without logic or rationality, and of little consequence. In this mildly skeptical view, the *feng shui* practitioners are just simpletons. Still, the naysayers who summarily dismiss the flow of energy as nonsense may want to visit a classical garden in China, practice T'ai Chi Chuan, Qi Gong, or another of the martial arts, join a drummer's circle, or surf an ocean swell, before finally coming to a conclusion about it.

If it's hard to accept the tenets of *feng shui* at face value, then one might simply imagine how a location is linked to the planet's elements. This, too, partly gives a place its auspicious personality. Take my situation for example. The physical setting of our home in Wolf Gap Holler connects it to the surrounding hills (*earth*) and flowing streams (*water*). The sun (*fire*) warms the house and nourishes the plants that grow in the gardens as well as the algae in the pond. Our home is set in a forest (*wood*) at the base of a ridge that protects us from the wind. Above our home is the sheltering sky. We keep warm in the winter by burning wood in a stove. The rocks in the garden contain minerals from which (*metal*) is derived. Beyond all that, though, is a point of view. At the very least, the alignments of a *feng shui* site make it an attractive place to relax. After all, who doesn't enjoy sitting against a rock warmed by the sun, situated in the shade of a tree at the base of a mountain, looking out upon a body of water? ❖

CHAPTER SIX

The decision some people make to live a simple life helps bring into alignment, however minutely, the global imbalances in resource consumption. It also allows a person to reach into the past, bring forward traditional wisdoms about Earth, and keep them fresh and alive. To live more simply is necessary not only for planetary balance but for a satisfied personal life.

A Scale of Life

My goldfish pond is a memento of the weight of the snow from a winter storm that swept through Wolf Gap Holler, and it provides me with a vantage from which to observe the world. Nature has rebounded in the nearby landscape; without a direct memory of it, a person might not know that bad weather had once visited the valley. As I settle deeper into my home and seek ways to live that are at least cognizant of the natural world, I am reminded that recovery is as much a part of life as devastation, and that the goal is to not impose such a heavy burden on the planet that it cannot hope to heal itself. In small ways we seek this. Jennifer and I mulch heavily in the gardens to suppress the growth of weeds without using chemical herbicides. We apply organic fertilizers to nourish soil and plants, mainly using truckloads of horse manure obtained from a neighbor's farm on Red Lick Creek. We heat our house with a woodstove, hang our laundry out to dry in the sun on clear days, and try to remember to turn off the lights when we leave a room. We grow some of our food in the garden.

None of this means I live self-sufficiently or without impact on the planet. My carbon footprint is evident whenever I turn around to look where I've been. This is exponentially true each time I fly to a distant place on the planet. My contributions to carbon emissions on international flights erase almost everything I might do at home in respect to the carbon cycle. It is a great irony of my life. The vegetables and fruit we grow in our garden seasonally augment our meals, but in no way provide the bulk of our food. We use electric heat when we leave home in the cold season and air conditioning on especially hot and humid summer days. I dig trees and ferns from the forest and transplant them in my gardens, but I also grow propagated specimens that originate in far-off greenhouses. I commute to work in my truck. My life is a web of entanglements that radiates outward from the goldfish pond like the dew-laden designs of a garden arachnid, connecting me to the world and its natural resources and all the complicated events of modern society. In the end, I'm a global person trying to live thoughtfully in the magical light of a particular place on the planet.

Many folks living in Wolf Gap Holler and on Red Lick Creek are much better suited to a natural life than me. Some have special skills and aptitudes that make it possible. A few even seek fully autonomous lifestyles, an admirable goal but mainly impossible to achieve nowadays. Even those living off the grid are linked to the global economy through its systems of information, power, markets, infrastructures, and community. In frontier times, everybody living on Red Lick and in Wolf Gap Holler was self-sufficient because they had no choice. At a distance of twenty miles, the valley in 1775 was the furthermost outpost of Fort Boonesborough, one of the first European settlements west of the Appalachian Mountains. The Red Lick settlers were on their own, with only one another to turn to in times of need or distress. Some people emulate a frontier lifestyle on Red Lick, but it's a personal choice now rather than an inherited necessity.

Tucked away in tiny Red Lick hollows are descendants of the valley's original pioneers. Some follow a more or less traditional way of life. Other more recent arrivals also attempt to follow in the furrows of the frontiersmen. Their reasons are multifold: personal freedom, suspicion of government, tradition, belief in God, fear of an impending apocalypse, a desire to be healthy, an Elysium dream, poverty, touching ancestral roots, invoking planetary limits. In any event, the frontier ethic, which enjoys an iconic place in American history, has a long history in Red Lick Valley, but the self-reliant path has become, in its own way, exclusive and limited to only a handful of people. That is because there's simply not enough arable land for everyone who might like to live independently to actually do so. And those who do have sufficient amounts of it closely safeguard their landholdings. This all makes an off-the-grid lifestyle its own gated community.

Many community efforts at self-sufficiency embrace like-mindedness among its residents and thus sometimes suffer the challenges of being a commune. Wolf Gap Holler, even in its hippie halcyon, was never more than a collection of neighbors. While some of my neighbors try to live off the grid, it's not as participants of a social movement but rather because they prefer a certain lifestyle. Most aren't interested in living in a commune. Almost everyone who strives toward self-reliance, however, finds they cannot really live alone because the way forward requires a step back to a time when hard work, good neighborliness, and an abiding sense of natural locality were recognized as essential ingredients of anyone's life, regardless of where or how one lives.

An hour's drive southwest of Wolf Gap Holler takes me to one of America's most celebrated sites of an intentionally designed natural community. Shaker Village of Pleasant Hill was established in 1805 when there was little national infrastructure to speak of, certainly not in that part of Kentucky, and off-the-grid living was the only way most people *could* live. The settlement occupies three thousand acres of rolling hills and stone fences along the banks of the Shawnee Run near the Kentucky River. It was started by a small group of Shakers who had come to the region from New York for a revival called the Second Great Awakening and decided to stay. By 1820 the village registered more than five hundred residents. For several decades, the settlement was stable, with people coming and going, but it slid into decline by mid-century. In the 1880s, the village held only a few widows with small children and a handful of indigent men. By 1910, it was pretty much over. The last Shaker resident, Sister Mary Settles, died in 1923.

A hamlet of handsome buildings marks the center of the old Shaker village, which is now a national historical park and popular tourist destination. The structures have been restored and the place looks productive—all neat and tidy and well-maintained—but nobody actually lives there: the Shakers sealed their fate with their vows of celibacy. The brick and clapboard buildings are still surrounded by fields of grain, potatoes, orchards, and vegetable gardens, so that it remains a working farm, not in a self-sufficient way but rather as an educational experiment. In the Shaker days, a division of labor was mainly based on gender. Men made brooms and coopers' ware, raised livestock, and tanned animal skins, while women wove textiles, processed cheese, and canned fruit preserves. They designed their own water system, bred cattle, and made herbal medicines. Agricultural produce was shipped by river barge along the Kentucky, Ohio, and Mississippi rivers to New Orleans. For a century, the Pleasant Hill Shakers grew most of what they ate, hunted, built communal homes and meeting houses, invented tools, and bartered with the outside world for the limited odds and ends they couldn't produce with their own hands.

A stroll through the site today provides a glimpse into its self-reliant past. Discipline governed Shaker society, and while life was natural, it wasn't always free. A harsh economy of space and the fine craftsmanship for which the Shakers are world-famous reflect the tight control they exercised on whimsy and flights of the imagination. Wandering around the hamlet, peering into the old buildings, and reading about Shaker customs, I wondered at such a lifestyle. Their society was filled with all manner of prohibitions, which must have been burdensome for some of them, and their spirits seemed to have been shackled to a rather hellish concept of the afterlife. Nowadays, the contract workers who re-enact Shaker life at Pleasant Hill for tourists wear pleasant smiles and period costumes while demonstrating the early survival skills. The settlement is a living museum, rooted in precepts for living naturally in the world, but not a model for a sustainable society in modern times. Its example was truncated by a changing world and its own restrictive ideology.

A few hours by car south of Wolf Gap is another well-known intentional community: The Farm. It was established in 1970 by a caravan of young people who fled San Francisco for the hills of Appalachia. For a couple of decades its residents led a self-sufficient life, working together and holding possessions in common. The Farm peaked in 1980 when more than a thousand persons lived on its 1750 acres. They raised their own food, made a soybean-based frozen confectionery called "Ice-Bean," hosted a resident rock band called "The Farm Band," which recorded albums with names like *Up In Your Thing*, grew marijuana—which landed some Farm residents in jail—enjoyed group marriages, and quietly initiated many good works for which they are little remembered. When asked about The Farm, most local people can only recall the pot, long hair, and promiscuity.

Today about two hundred people still live on The Farm, but they don't live *off* the farm. They commute to salaried jobs and live in homes outfitted with electricity and running water. Kitchen gardens provide seasonal produce, but The Farm is not self-sufficient in food. Curious people are welcome at The Farm, but visitors need permission to enter the grounds, much like Wolf Gap Holler before Matt died, when he stood by the side of the road vetting suspicious incomers with a piercing look and a pistol at his side. The Farm is private property. People residing there continue to contribute charitable work to the local region and, increasingly, to the world. One of its founders, Ina May Gaskin, won the Right Livelihood Award for her work in midwifery and natural childbirth. She received the prize in Stockholm alongside other recipients hailing from China and Ghana. While still a practice in nature-based living, The Farm no longer provides a place where a person might live separate from the rest of world.

A contemporary of The Farm is Turtle Island Preserve, located across the Eastern Divide from Wolf Gap Holler, among the Blue Ridge Mountains of North Carolina. The preserve was established by Eustace Conway, survivalist *par excellence*, as a place for people to learn how to live in the wilderness. Mr. Conway is the subject of a book by Elizabeth Gilbert called *The Last American Man*, which makes clear that he is no average person. Mr. Conway once rode a horse across America by living off the land. He hiked the Appalachian Trail in the winter with no support, not even a granola bar. He's steeped in the ancient lore of wild foods and forest medicines, highly-focused, makes his own primitive tools for hunting and dressing game, and is known to occasionally perform minor surgery on his body. Mr. Conway attracted a fringe of survivalist-types who might have felt quite at home in Fort Boonesborough, and together they carved the Turtle Island Preserve out of a chunk of rugged Appalachian mountainside. The personality of the preserve, though, has changed with time. Mr. Conway discovered that most people don't share his intensity of purpose. Visitors to Turtle Island can learn important wilderness skills, but the preserve has shifted away from being a place for people to live *in* the wild to a center for people to learn *about* the wild. It, too, has become something of an ecological tourist attraction. Turtle Island developed its own website and Eustace Conway gained a starring role in his own television reality show called "Mountain Men."

In India, half way around the globe from these American communities, is a hardscrabble landscape filled with impoverished tribes and dirt-poor farms. The tiny settlement of Rasulia was established there in the 1930s by the Quaker Society as a self-sufficient village based on ideas of social justice and natural living. Unlike its Shaker counterpart, the village has no fancy architectures or neatly built stone walls. It was a place neither for wilderness survivalists nor aspiring hippies, but simply for some villagers who wanted to live outside the limits of their caste-bound lives and among others in a socially free and quasi-natural regard. In the 1980s, its charismatic leader, Partap Aggarwal, adopted the organic principles espoused by the Japanese farmer-philosopher Masanobu Fukuoku, author of the highly-acclaimed little book *One-Straw Revolution*, and Rasulia soon became a model agrarian community. Fukuoku's quiet study in Zen and farming became a best-selling handbook in the 1970s for back-to-land practitioners in wealthy America, and Mr. Aggarwal adopted it as a guide for Indian farmers living in one of the poorest places on the planet.

Rasulia was in decline when I visited it in 1985. It took several days of hard travel across central India on rickety buses and local trains to reach the village. I rode the final few miles in a bullock cart through a midnight landscape of dusty roads, cicadas, and shuttered homes. The clip-clop of the oxen team and the creaking wooden wheels of the wagon awoke people whenever we passed a residence. Someone called out suspiciously, asking who was there at that late hour, and I replied that I was looking for a friend and gave his name. The conversation went back and forth in the darkness until a boy sleepily came out and guided me to my friend's thatched bungalow.

I spent a week in Rasulia catching up with my friend and observing life there. I was living at the time in a mountain village in Nepal and was accustomed to the sort of life that goes on in a tiny rural community in that part of the world. I found in Rasulia a departure from conventional agrarian ways. Its residents had embarked on some novel approaches to farming and social organization. They cultivated the soil using a no-till method. Fields were held in common rather than as private property. Decisions required a consensus and everyone worked side by side on community projects regardless of caste or ethnicity. A shared methane generating plant processed livestock manure into cooking gas and organic compost. The gas was piped into homes, allowing people to prepare food on clean-burning stoves rather than on the traditional smoky dung files that are commonplace in rural South Asia. The residual manure, meanwhile, which is rich in nutrients, provided them with an odorless field fertilizer. The villagers collectively owned a loom house and wove a handspun fabric called *khadi* from the cotton they grew in their fields. It's a simple cloth that was worn by Gandhi during his struggle for India's independence from British colonial rule, and the material became symbolic of the struggle of local economies against global corporations. For the Rasulia farmers, it highlighted their struggles to free themselves from the repressive aspects of Indian society and the pernicious effects of globalization. I joined the villagers on some evenings when they met to discuss farm work and community affairs, and observed how people could freely voice their concerns and ideas.

It wasn't clear to me, though, how Rasulia might serve as a model for other places in the world. Minor tensions were simmering in the village. Its admirable egalitarian strategies would be hard to implement elsewhere in conservative, caste-based India. Not long after my visit, Mr. Aggarwal left for Bangalore, and I later learned that Rasulia had quickly suffered a lack of cohesion and shared vision. It had been kept financially afloat in hard times by the international Quaker organization and so benefited from outside intervention in ways that other places in India did not. It's not clear that its guiding principles of organic farming had much chance elsewhere in India where industrial-scale farming is thought to be necessary in order to keep food production at a pace with a spiraling population. Committed individuals continue to fight for organic lifestyles in India but too often find that entrenched politics lock people into cycles of dependency and vulnerability. In this way, Rasulia is not so different from other places in the world that aspire to a sustainable society.

The questions surrounding the design of a natural life are perplexing at all scales, more than George and I might even imagine when we pose them on some mornings at my goldfish pond in Wolf Gap Holler. The topic of sustainability is found in the curricula of just about every school in America. Entire books and conferences around the world are devoted to it. Agencies of international development organize sectors of the world economy around it. All conceivable ways of framing it appear in the public discourse, from national constitutions to refrigerator magnets. The future of the world revolves around the notion of a sustainable life—it is that big. Though its scale lies beyond our comprehension, George and I still like to talk about what it means to live a sustainable life. Solving seemingly intractable world problems is one of the things we enjoy doing when we sit quietly with a cup of coffee in the morning in my Wolf Gap gardens. And as big an idea as it is, sustainability may in the end simply boil down to a person living thoughtfully in a place of her or his choosing.

The planet hosts more than seven billion persons, and at current rates of growth, our global population will exceed ten billion by 2050. Most people in the world are poor, landless, and live in huge cities without access to proper diets, water, or health care of any kind. A great many suffer from civil war, political dysfunction, drought, famine, or natural hazards. Living in bleak circumstances, the world's poor consume only a fraction of the energy and resources that a typical American requires to live what is considered to be a normal life. But what *exactly* is a normal life when it comes to what we need and want? Most people in

Western societies consume planetary resources at unsustainable rates not because we choose to be rapacious or greedy but simply because we have learned no other way. We believe we are entitled to such a life. We have come to learn it is the *good life*, a *normal* life. Consider that against this: if the world's poor actually were to achieve a living standard comparable to that of an average American by emulating the Western models of progress and consumption, no one would celebrate because the carrying capacity of the planet would collapse, and we all would suffer. Earth cannot sustain that level of demand. In other words, I live the way I do in America *because* other people living elsewhere in the world do not. Of all the troubling facts of my life, this is one that looms large and seems most intractable.

The decision some people make to live a simple life helps bring into alignment, however minutely, the global imbalances in resource consumption. It also allows a person to reach into the past, bring forward traditional wisdoms about Earth, and keep them fresh and alive. To live more simply is necessary not only for planetary balance but for a satisfied personal life. But if such a lifestyle requires enough space so as not to bump into one another and enough freedom to roam around in private forests and to till many acres of farmland, then what hope does it offer at a planetary level? A lifestyle that is open to only a handful of people isn't much of a solution to the global problems of world hunger, poverty, or ecological collapse. The numbers simply don't add up. What can one say to an orphaned child in a third world city who lives on the street and begs for a life … buy forty acres and a mule?

It takes a lot of land to produce enough food, timber, and firewood to support even a small handful of true off-the-grid types. The ones who live near me tend to have access to large tracts of it. What promise, then, can such a lifestyle possibly hold for the world's poor, who number in the billions, are mainly city dwellers, and have no place else to go? I think about this question a lot, and George and I occasionally discuss it in the garden. I built a career around it in my early work as a geographer, taking the problem to some of the most remote and destitute places on the planet. I'm sorry to report, after all that time and effort, I don't know much about what will work in the end. There is no silver bullet. During one of my recent visits to India, on the day I was to catch a return flight home, I took an early morning stroll through the streets of an old part of Delhi. It's something I like to do because it puts me in touch with a certain kind of life. I was feeling impatient with India, frustrated, and a bit on edge, which always signaled to me that it was time to go home, and so I wanted to connect with the place before leaving it. A winter fog blanketed the city in cold dampness. I walked past a group of street kids huddled around a fire made from a bundle of tightly-wadded plastic bags. The black smoke was heavy and noxious as it rolled down the alleyways. I walked over to them and asked what they were burning to keep warm. The youth understood that I already knew, that I was just making a point, and caught the opprobrium in my voice. One of them looked at me and snapped, "*Oil — just like you*." The remark stopped me in my tracks. He was right, of course, because plastic bags are made of processed petroleum. I grew embarrassed and walked away.

If there is no single model of a natural life, then what standards can be applied to a life that is designed with nature and at a proper scale? Jennifer and I try to live a thoughtful life in Wolf Gap, and that desire extends to the land we cultivate, which isn't much really, certainly not enough to live anything close to a self-sufficient life. We grow some organic vegetables and fruit, tend the gardens, and tinker around in the woods. But what does that effort mean? Mainly, it gives us a chance to learn about the natural world, to center ourselves to some extent within it, and this, in turn, leads to choices that we might not otherwise have to make. It extends our thinking beyond ourselves and provides us with a chance to get *outside* of ourselves. There's a lot of attention placed on finding one's "inner self." I think mindfulness is to be found mainly outdoors, alive in the moment, and in the company of good people.

How is it possible for me to understand who I am without knowing how I fit into the world? The need for natural interaction seems as true for city dwellers as it does for country folk, which is, I assume, why a smart city has a lot of parks. They provide places in cities where people can learn about nature. We'll need many more urban parks in the future. The majority of the human population now lives in towns and cities, and by 2030 more than two-thirds of humanity will be city-dwelling. The 2010 global census showed that we crossed the threshold from being a rural species in the past to an urban one in the future. There is no going back. It is equally clear that our rural landscapes must sustainably support not only those living in them but also urban people throughout the world. Any solution to this impending conundrum will require not only new agriculture and energy models, but entirely new ways of thinking about resources and lifestyles.

I know that I am fortunate. I don't have to visit a city park to see how I fit into things; I *live* in a park, and I'm not alone. I reside among a community of fellow park dwellers strung along Wolf Gap Road like prayer beads on a rosary. I don't live in a commune or anything like that. We're neighbors. A lot of us don't always see eye to eye on things. We have different histories, beliefs, and incomes. We live among forests and gardens, attend to the details of country life, and try to get along with one another as best we can. We try to live a rural life while also juggling the demands of being members of contemporary American society. We're not unique in these ways, but given the global demographics, our rural numbers are dwindling.

As I look around America and travel to various parts of the world, I see city life ramping up. Moving so fast, it's hard to find time for others, nature, or even ourselves. Increasingly estranged from the natural world and, I suspect, from the communal spirit that resides deep within us, yet crowded into ever more congested spaces, how do we continue to move forward together? It seems unlikely that a solution will be found in yet another round of technological advances, which simply allows us to further distance ourselves from the world around us and from direct contact with one another, caught inside an increasingly virtual matrix of life that for some people has become a substitute for reality and for others a mockery of it. To live a natural life, however modest and constrained, would seem to require being in the continuing presence of natural forces, anchored to a place, living moment by moment, and in kind consideration of others. ❖

CHAPTER SEVEN

Scaled down, the most important places are those that are close to home, where we spend most of our time, and to which we have our strongest emotional attachments. When we enlarge our mental maps to take into account far-away lands, so that a greater number of places become known to us in a thoughtful fashion, we increase our capacity to feel a personal commitment to an ever-expanding conception of the world. We become empathetic with Earth and its myriad peoples.

A Sense of Belonging

It's natural for us to link our well-being to the human community as well as the place in which we live. These ties bind us to one another and help us find our way in the world. It is the underlying message of sustainable communities: we need one another just as much as we do nature. While my goldfish pond is a natural ecosystem, I can't *live* in my pond. I just sit beside it on some mornings with George and drink coffee and feed the fish. Human life, it seems, must be lived in a *human* ecosystem, one that weaves our societies into the fabric of a humane world.

Life in Wolf Gap Holler is a microcosm of such ideas. It is also a prosaic life. We work together to cut and split a winter's supply of firewood or to build a house. We exchange plants and gardening tips and share food at neighborhood meals. We borrow tools from one another and help out when asked to do so or if we notice that it's necessary. Some of us take walks together on paths in the forest or along Wolf Gap Road in the morning. We pick up trash that gets tossed to the side of the road by thoughtless motorists. Some of my Wolf Gap neighbors make useful things; potters, weavers, and wood-workers live in the valley, and their efforts find their way to some of the rest of us. We join forces to clear the wreckage of natural events like windstorms, blizzards, or floods, and we celebrate solstices and equinoxes with parties beside a bonfire. We are a community of neighbors in such ways.

I've resided in different communities during my life, but the Wolf Gap Holler community may be the first I've lived in with a conviction to stick around. That doesn't make it an intentional community, just a preferred place for me to live. Sometimes I think I just *happened* into the others. I was born and raised in Michigan in a blue-collar town situated on a Great Lake, on a street where neighbors knew one another and the kids all played together. It was a quiet, ordinary neighborhood, a good place to raise a child, and I'm the beneficiary of that. It was my first community, but not as a matter of choice because I was born there. I can't take responsibility for being there, just for leaving. Still, if I were to choose a community for my childhood, I can't imagine one much better than the one in which I was reared; my hometown's slogan is Blue Water Wonderland.

I left my childhood neighborhood to go to college in a distant city where I lived in a house near campus with a bunch of other students. You can imagine what *that* was like. It was hardly a community. More like a rugby scrum. After graduation and a certain amount of roaming around the world, I moved to Hawaii to pursue graduate study at the East-West Center. I occupied a small room in a dormitory alongside two hundred other students from around the world, and we all got along pretty well and learned quite a lot from one another. The potluck dinners we had at the East-West Center were some of the finest culinary experiences I've encountered in the world, not because we were especially good cooks—we were just students—but because we ate food from all over the world that was prepared from the heart: baked taro and fish concoctions from Polynesia, coconut curries from Thailand, beer from Australia, kebabs from Kabul, and so forth. We regularly met in the evenings in our communal kitchen, cooked our food, and sat down together for a United Nations-style meal.

Living in a tiny room on the top floor of a dormitory was hardly living in a community in any natural way, even though my window opened onto the Pacific Ocean and across the hall was a view of tropical green mountains. Beyond the East-West Center was greater Hawaii: emerald forests and waterfalls, sunshine, and sapphire ocean waves. Being a Caucasian man from Michigan, the community of Polynesians that comprise the native population of Hawaii was mainly out-of-bounds for me. Most of my friends in Hawaii, like me, were white outcastes called *haoles*, and even if our sympathies lay with native rights, we were also colonialists and often were treated like ones. Still, I had a great time living in Hawaii, eventually finding my way into the wild heart of the islands, and the castaway life proved to be a good one in my experience.

In a break from oceanic life, I went to Nepal and lived for a year in a mountain village where sense of community was ageless and people lived entirely off the land. It was no American-style back-to-the-land community. There was no option in the Himalaya for villagers to live other than self-sufficiently; a person had to walk for many days before coming upon a road or electricity. The only running water was that which flowed down rivers. I was in Nepal to study subsistence life for my Ph.D. thesis. I remember arriving in the village after climbing for a week on mountain trails and meeting with the community elders to explain my purpose and to ask if it was okay for me to stay in their village and observe their lives. When I look back upon it, I wonder that I had the gumption to bring it up; imagine, a young foreigner appearing out of the blue and petitioning villagers to be housed and fed for a year while pestering them with a lot of questions for the purpose of an obscure academic study. Anyway, it was very gracious of them to allow me to stay, and it worked out splendidly. I paid a small sum for a place to sleep and two simple meals a day. I participated in as much of village life as I could manage. I got along well with most of the villagers, was a friend to some of them, and learned a great deal about their life and my own. Still, I never was a *part* of that community, although I did live a more or less natural life. I ate what was grown and raised in the village, slept on a wooden floor above a livestock pen where the breath of the animals kept me warm at night, read by candlelight, bathed and washed my clothes in a nearby stream, drew water from a spring, and shat in the fields at dawn just as the sun rose and the icy peaks took on a warm, rosy-pink glow.

I returned to Hawaii after my time in Nepal to write my dissertation, and there met my first wife. She was a Polynesian woman, the daughter of a Samoan chief whose lineage could be traced back for three thousand years to a warrior island called Manono. We tried living in Samoa for a while—until I couldn't take it anymore. That might sound strange. Samoa is the picture of Polynesia: coconut palms draped over blue lagoons, lingering sunsets, and on most days the cool promise of a trade breeze. It all conjures up an easy life in the islands. But such a picture-postcard view doesn't tell the whole story of life in Polynesia. The image of a coconut tree leaning over a beach, for instance, doesn't explain the fact that saltwater is blowing against the tree's trunk, causing a cellular collapse in its bark, pulling the tree toward the water as it grows. One side of it winds up being shorter than the other. The palm tree is not a dreamy fashion model on a South Seas photo shoot. It just *looks* like it wants to drape itself over the glassy lagoon; in fact, it's being twisted out of all normal proportions and just trying to stay alive. Coconut palms have genetically evolved, though, to live alone on a salt-blown beach and can tolerate quite well the changes such an environment might induce in their morphology. People are different. I was being asked in Samoa to be someone other than myself.

It was a big adventure for me, though, living in Polynesia and traveling to islands that lie scattered like velvet buckshot across the ocean. I sailed on trading vessels to far-flung coral atolls, romped through pristine rainforests that botanists had yet to survey, and scrambled up steaming volcanoes. I fished and swam among some of the most pristine coral reefs in the world. Dividing my time between a coconut plantation in the mountainous interior of Upolu Island and a thatch beach bungalow on Manono Island, both delightful places, I enjoyed an inside track to Polynesian culture. I would drive my father-in-law's truck to village ceremonies where he officiated as the Paramount Chief and haul loads of baked taro, bananas, fire-roasted pig, and mountains of fresh fish to feasts that made our East-West Center potlucks in Honolulu look puny by comparison. Off-times, I frolicked in freshwater pools at the base of waterfalls, hiked through native jungle on islands that were so small and remote they didn't appear on any maps, picked mangoes and papaya where they grew wild and ate them fresh from the tree with their tropical juices dripping down my arms. At times, I felt like a real nature boy in Polynesia.

But I never really felt at home in Samoa, despite my ties to the place through marriage, not even among my former wife's extended family who graciously welcomed me into their affairs. A person either is born or not into Polynesian society, and if not, remains a visitor for life. Samoa is a lot like Kentucky in that regard. There simply is no way for someone from the outside to wriggle into such a society. It's an understandable exclusion in Polynesia, where people are joined by thousands of years of blood line and cultural inheritance and isolated from the rest of the world by the Pacific Ocean, but less so in Kentucky where I came to live in Wolf Gap Holler. In the end, my Samoan marriage didn't last—in part, due to the strong tug of time and space: I couldn't remain in Samoa because I saw no future for me there, and she couldn't leave because of the strong pull of her cultural past. Any way it might have turned out, one of us would have been a fish out of water for life.

When I moved to Wolf Gap Holler in Kentucky, it was with the idea of living in the countryside as part of a rural neighborhood. I wasn't looking for an intentional community, just a pleasant place to live with plenty of room and some nice people nearby, and without a lot of restrictions on what I could do. From the onset, I found no pressure to fit into Wolf Gap society. Most of my neighbors are from somewhere else, and so we had that much in common from the beginning. We try to look out for one another in some ways; for example, when a person gets sick and needs support or when shared labor makes better sense than going it alone. We have fun together and learn things from one another. That's about as much of a community, though, as most of us could imagine being a part of.

I've always had a roaming nature, which is why the notion of a settled community life had been an anachronism for me in the past. I think I've always valued the idea of it, but never made much of a commitment to belonging to a place for very long. I left the town where I was born as a teenager to embark on a backpacking trip around the world and have been traveling regularly ever since, having managed along the way to translate my wanderings into a career as a geographer. Given my kind of work, I rarely feel like I have a job to go to in the mornings. And when neither on campus nor off traveling in the world, I have Wolf Gap Holler all around me. My journeys relieve some of the impatience I feel when I stay put for too long in a single spot. I'm nomadic by inclination, and so a settled life in Wolf Gap is a continuous work in progress.

Ironically, I've discovered that one of the key ingredients to belonging to a community is the practice of patience. If we live in a community, which is a type of relationship, and among other people, we can't expect everything to always go our way or at our own pace. The need for patience might seem obvious, but it's not, and for good reason. Patience is in short supply these days, and not just with me. Impatience, it seems, has many modern triggers, including microsecond delays in the operations of a computer or mobile device. What prospect is there, then, for the exercise of patience necessary in the long run of life for a place-based community?

I'm not sure one can learn to be patient if not born into it by personality, but I think we can all keep working on it. Impatient adults living in a neighborhood, like bullies in a schoolyard, can lead to intolerance, grudges, boundary disputes, and even violent actions. Living in a community, even a loose one like Wolf Gap Holler, people don't always see eye to eye on things or come to the same conclusions at the same time. Scaled up to larger society, we can't avoid the differences that exist among us, but we can try to be patient with them. A community without the practice of patience among its members has little chance of remaining intact. It will fray at the edges and then, getting close to the heart of things, despair, and alienation can set in. Communities, and the larger societies of which they are a part, begin to disintegrate. People leave or turn aggressive when there is no tolerance in society.

What I have learned most about patience, though, is that it's a relative thing. We don't all share the same degree of *sang froid.* And we have different thresholds for different kinds of things. Some gardeners, for instance, exhibit a great deal of patience whenever they plant a tree; after all, it will mature over a lifetime. Other people like to grow annuals that bloom overnight. I've found bamboo to be a good plant for impatient gardeners. A Yellow Groove variety (*Phyllostachys aureosulcata*) that I grow in a small grove at the edge of my Oriental garden sends out new culms in the spring and within a week they're taller than me. They attain their full height in a single season. Some species grow three feet in a single day. Bamboo grows so fast, in fact, that you can actually *hear* it grow. It's quite remarkable. When the new culms in my garden are head-high, I sometimes go into the grove at night to listen to the growing bamboo. It makes a soft, dry rustling sound, as if it were stretching itself out, coming into its own fullness. Not only does a bamboo culm grow quickly, but the entire plant rhizome spreads underground remarkably fast. The temperate varieties that can be grown in Kentucky are called "running bamboo," and for good reason. They'll quickly outrun their boundaries. Although I love bamboo and plant it widely, I would never recommend it as a garden specimen for someone living in a town or city. Planting bamboo as a boundary screen in a subdivision cul-de-sac would not make for good and patient neighbors.

Now and then, George and I practice a sort of patience when we sit by the goldfish pond. If some person were to walk by unbeknownst to us, they might think from the looks of it that we are doing nothing at all, that we are idlers. In fact, though, we're quietly biding time, waiting for some tiny action to occur in the pond: the rise of a goldfish, the croak of a frog, or just a thought to arise from within and punctuate the stillness of the pond's surface. One of the rewards of patience is insight. When George and I sit by the pond, it just *seems* like we are doing nothing, but we're actually practicing patience with the idea, I suppose, of one day becoming proficient at it. George calls this diffuse sort of waiting with no particular outcome in mind, just a general wash across one's being, "atmospheric patience." By this I think he means that we all have the ability to surrender ourselves to a moment and gain a sense of grace that underlies the practice of mindfulness. It seems to me that atmospheric patience is akin to deep meditation.

My goldfish pond is composed of many residents: fish, salamanders, frogs, plants, and insects, all co-existing if not peacefully then at least somewhat patiently together. The pond's frogs are Zen masters of tranquility, maintaining a stillness that is, quite frankly, unnerving. The pond and its residents compose an ecological habitat. My social habitat, meanwhile, is composed of lovely hills and valleys, cultivated fields and gardens, and people's homes. It's not a habitat that is as well-delineated as my pond because of the facts of modern life: we have no instinctual boundaries anymore, just lines on a map. Moreover, we're not *stuck* in Wolf Gap Holler like the goldfish are in my pond. At any moment I can get in my truck and drive away. Moreover, the scale of our lives is global. We consume resources, discharge waste, communicate, and participate in social policies that extend our reach to the furthest corners of the planet. Earth is the habitat in which I live. This means that the lives of distant people whom I may not know have come to depend upon how I live in Wolf Gap Holler. This interdependence calls attention to the fact that truly our home on Earth spans its antipodes and hemispheres.

Such an enlarged sense of life is one of the reasons I travel. It provides me with a direct apprehension of how I belong to a planetary ecosystem and global human community. The denizens of my goldfish pond know their world very well. They know its perimeters and hiding places, where food is available, when the water is low or of poor quality, and the places where they are most vulnerable to predators. The pond, though, is a small place and it's reasonable to assume that one of its inhabitants might learn all about it. Earth, on the other hand, is so large that it's impossible for us to know much of it very well at all. Yet, in learning more about the world, we can turn its abstract locations into more meaningful places. This transformation may be one of the most promising aspects of globalization, for as distant localities take on a greater meaning for us, we tend also to accept more responsibility for them.

Scaled down, the most important places are those that are close to home, where we spend most of our time, and to which we have our strongest emotional attachments. When we enlarge our mental maps to take into account far-away lands, so that a greater number of places become known to us in a thoughtful fashion, we increase our capacity to feel a personal commitment to an ever-expanding conception of the world. We become empathetic with Earth and its myriad peoples. If the majority of us were to actually do this, that is, consciously expand the scope of our lives by considering the true scale of it, and subsequently take responsibility for the greater world in which we live, it stands to reason that the planet's environment would benefit from our stewardship of it; distant peoples would learn of our respect for them, our local communities would become more sustainable places for us to live, and we would all be less in harm's way and much more likely to live a satisfied life. ❖

CHAPTER EIGHT

Many people realize that we are pushing nature beyond reasonable limits, and as a result, losing some of our capacity for human goodness; for surely, the two — nature and humanity — are inseparable. The expanding global economy is stretching the carrying capacity of Earth, and we are tasked with learning again how to live a sustainable life. It is the most important project of our lives.

The Control of Nature

The word ecology comes from the Greek terms *oikos* and *logia*, which taken together means "study of household." The idea suggests that all living organisms are dependent upon one another and the natural setting. Ecology operates at all planetary scales, from the tiniest creature deep in an ocean cave to the firmament. The German naturalist Ernest Haeckel coined the term in 1866 in support of emerging theories about evolution and to explain how energy and matter flow through the world. Haeckel's thoughts about life were wide-ranging; he was a philosopher, artist, and physician, as well as physical scientist, and over his lifetime he named many hundreds of plant and animal species, studied organisms ranging from protozoa to early hominids, wrote scientific treatises as well as popular travel books, and painted valued oil canvases of natural landscapes and biological specimens. Haeckel's eclectic approach to the study of natural history matches the range of topics George and I might entertain on any given morning sitting beside my goldfish pond. We shift tack in our conversations in a heartbeat. Like sailors adrift in an equatorial sea, our thoughts fluff of a sudden as if we were in the midst of the doldrums. We imagine ourselves to be lateral thinkers when others might say we just lack focus. An idea like Haeckel's "ecology," in which all organisms clearly are belayed to others, is a lifeboat when the water in the goldfish pond gets murky. It helps us see more clearly how our lives naturally fit into Earth's rhythms.

Haeckel's early work in ecology centered on his study of turtles, which, coincidentally, sometimes appear in my goldfish pond. A large snapping turtle (*Chelydra serpentina*) once took up residence in the goldfish pond, bothered the fish, and ate the plants, taking especially large bites out of the lily pads, which must have tasted like exotic arugula to the cartilaginous creature given the way he went at them. The incident taught me something about food chains, which of course is a huge part of the way Haeckel envisioned an ecosystem to work. The snapping turtle was a natural visitor to the garden but its presence in the goldfish pond put its other residents at risk. To build a pond in a garden and watch it spring to action on a daily basis brings a window of clarity onto the often obfuscate interlacing of the natural world. I finally had to remove the turtle with a seine, though, because it proved to be such a ruffian in the pond.

An important early contributor to ecology was an English clergyman, Gilbert White, who in 1789 wrote *The Natural History of Selborne*, in which he recounted his weekend jaunts through his Hampshire parish, closely observing nature while also visiting his congregation. Mr. White had an easy, colloquial writing style that made his ideas about ecology accessible to a wide range of people. His book became famous in its day, a prototype for the kind of writing we now call a nature essay. White paid close attention to matters of soil, climate, and the succession of plants, and was keen to know about the interactions of living organisms: how species *got along together*. Gilbert White was a religious person and imagined a sacred world in which humans coexisted with nature rather than living separate from it. He thought a divine energy emanated from all beings and flowed through the natural environment. He freely intermingled notions about faith and science, equating energy flow with a creative "life force," and wrote compassionately and with empathy, always quick to mention God in the same sentence as Nature. Gilbert White tended to see a thing in its entirety rather than narrowly in terms of its bits and pieces. Some people thought he was a quack.

Mr. White lived in the latter part of the eighteenth century, during the "Age of Reason," and his ideas struck many people as the antithesis of rational thought. He was considered by most scientists to be a quaint and harmless man, an eccentric pastor of a rural parish who obviously spent too much time outdoors, not someone to turn to for answers to the pressing scientific questions of the day. Serious scholars avoided him because they thought he lacked objectivity. In an era when practical problems were to be solved with action and not contemplation, a time, after all, of empiricism, logic, and industrial reason, even philosophers dismissed White's ideas as being antithetical to the new scientific methods, shushing them as the idle thoughts of a dilettante. Gilbert White was hard-pressed in his lifetime to win over his contemporaries in science or even those engaged in nature-based humanistic endeavors. But he attracted a great many admirers down the line, including Henry David Thoreau, the archetype romantic naturalist, and fostered an intellectual progeny that included many transcendentalist poets, artists, and writers. For such people, Gilbert White's appeal lay in his belief that one could neither understand the natural world nor a person's place in it without first admitting that humans are part of nature; indeed, that nature is an inescapable part of humanity.

But England was on the money march in the eighteenth century, churning out consumer products while people got rich and pollution spewed into the sky, land, and water. It was the era of the Industrial Revolution. The great machinery of this new economy was harnessed not far from White's quaint Selborne parish, among the factory towns of Yorkshire, Lancashire, and Staffordshire. The prospect of great wealth was reflected in these industrial landscapes and fueled people's imagination, so that the quiet, pastoral overtures of Mr. White's bucolic prose seemed to many of them only a mild diversion—like a quiet stroll in the park and not a prescription for life. The attentions of people were given over to the pressing matters of material progress, as if wealth was all that mattered, and his book was discounted by some as a thing to be read merely as leisurely entertainment.

Even now, in the twenty-first century, and despite the technological innovations of the intervening centuries, much of the English countryside appears as lovely as Gilbert White had described it more than two hundred years ago in his book about Selborne. Its scenic preservation is testimony to the idea that nature might have an aesthetic value for many people. I once hired a canal boat to tour a small part of White's world, traveling in waterways dug by the early industrial magnates to ferry goods between seaports and inland factories. I stopped often on my little voyage, moored the boat alongside paths where mules had once pulled the barges, and walked amid stone-and-thatch villages and poppy meadowlands, thinking how familiar they might have looked to Mr. White. I learned that the canals, with their antiquated locks, pumping stations, and aqueducts, had been engineered to be a self-contained circulation system. It worked like a gigantic garden pond: water in the canals flowed downhill under gravity and upslope through a series of locks, pulsing through the countryside with a metallic heartbeat of capitalism. The water went round and round, flushing England with the swollen arteries of newfound wealth. Here was a fine example of how society came to harness one of the key components of an ecosystem, the flow of water, in order to put nature to work making boatloads of money. Puttering about in my little vessel, adrift in a limpid metaphor for the Industrial Revolution, I followed in the wake of the early entrepreneurs who had sought to master nature for the main purpose of exacting from it great wealth.

The Englishmen abandoned their canals when railroads and highways proved to be more profitable at moving goods around the countryside, and the infrastructures remained neglected until the middle of the twentieth century, when Britain's tourism industry undertook a renovation of the waterways for recreational boating. The willowy canals then were dredged and locks repaired. Barges were stripped, painted, furnished with modern amenities, and decorated with colorful family crests: roses, clovers, hearts, or diamonds. In coming upon a fleet of the renovated boats lined up in an idle stretch of water, their hulls emblazoned with insignia, I saw them as an advancing flotilla of fancy playing cards. Some of the boats sprouted tiny potted gardens of flowers and herbs from their rooftops. White lace screened the windows of others. Swans idly swam by. Flanked by wildflower meadows and cottage gardens, my whimsical canal tour was rather like being afloat in a great water garden at English tea-time, ready for a game of whist. As if *I* were a small water creature in a goldfish pond and all of lush Victorian England was the garden.

My voyage drifted past quiet homes, sheep in pastures, pubs and parish churches, and fields of bright red poppies. The cruise went off without a hitch until one day I thought I'd turn the boat in a stretch of water without using one of the roundabouts that appear at regular intervals for the purpose. Jennifer warned me not to do it, thinking there wasn't sufficient width in the canal for the length of a spinning boat, but I didn't listen to her and got us stuck, instead, with the bow and stern of the boat jammed tight into the muddy banks. Oncoming vessels piled up in my imagination, as if I were blocking the shipping lanes of the Suez Canal, and we struggled to set our little boat free of the muck before someone might see us. I tried to appear composed all the while, as if in charge of matters and not simply another tourist having a bad go at being a skipper. Of course, there were no other boats bearing down upon us—it was only in my mind; we were alone on the canal at that time of day. I learned, though, how a goldfish might feel if it gets caught in the suction of the intake of my Wolf Gap pond's water filtration system: Would it be struggling for its life, *or just trying not to look like a fool?*

The pump I put in my goldfish pond to circulate and filter its water mirrors at a tiny scale the grand intentions of the English canals. They both are efforts to more fully control our world. We've always used our technologies and ingenuity to harness the planet's natural elements and to obtain from them what we need to be healthy, secure, and adequately fed, but the Industrial Revolution taught us to think beyond our needs and strive to obtain more of what we want. This opened the proverbial Pandora's Box. It's clear to most of us when we have what we need, but who is to say we have all that we desire? From whom do we learn what to covet? The research now is clear on this: our most influential teachers today are no longer our parents and families. Nor are they our peers and school teachers. Rather, the nexus of learning has shifted to the global media and its ancillary devices, funneled through sprawling information and financial networks, and with a primary instruction in mind: we need to increase our levels of consumption because the world economy and our own sense of self-satisfaction depend upon it.

The media message couldn't be more clear than if it were an advertising banner dangling from the stern of a dirigible hovering in the bright blue sky; we need to be more voracious in our appetites for the planet's resources. Therein, we are told, lies the source of our happiness and prosperity. It is the overwhelming message of modern times. Deep inside, though, we must know it is a false promise. In our acquiescence, in this handing over of responsibility to the global media and its multifarious commerce for helping our children learn who they will become in life, what to value, and hold to be true about life, we are delivering our children to the Captains of Industry just as surely as if we were living in bleaker times and handing them over to the Blackbirders in a West African slave port, bound and shackled for a tortuous journey to a New World plantation. We are taught to seek not only our happiness in ever-higher levels of consumption, but to find our very selves in it. What we buy is conflated with who we are as a person. We clothe ourselves in fashions that advertise the very companies that influence so much of our lives, and so we act as mere salary men peddling goods on a city sidewalk, all the while believing that others might think the better of us for it. We've thrown our hats into the ring and become the walking billboards of the global economy.

The industrial canals of England were small cogs in the machinery of the emerging global financial system, of which Britain was the undisputed master during the world's colonial era. The purpose of colonialism, after all, was to obtain resources and human labor at the lowest possible costs from wherever it was most readily available, bring the raw materials back to Europe's factory towns for production into consumer goods, and sell the manufactured products to people living across the globe. In this way, the traders became fabulously wealthy. England, of course, was not alone in doing this. Many countries in Western Europe and later America were involved in the project, for it proved to be grotesquely profitable, but the British were especially adept at it in the early years. It's an old story of plunder and economy taken from the pages of human civilization.

The rate and scale of resource extraction and industrial output as well as the toxic discharge of waste into the natural environment have accelerated greatly since the colonial era. Out of this has emerged a new global order that further divides us along the lines of wealth and power, promoting economic enterprises that require ever-greater amounts of energy and raw materials. Our societies have become places where a person seeks satisfaction in life through greater consumption. We in the West have become *uber*-consumers, and others in the developing world are soon to follow. The social system we inherited and have enlarged upon, though, is not sustainable for the planet or even for a single person's life. We are unlikely to keep pace with rising expectations, and the accelerated growth can only come at greater social and environmental costs. The planet has become overburdened by it all. Many people realize that we are pushing nature beyond reasonable limits, and as a result, losing some of our capacity for human goodness; for surely, the two—nature and humanity—are inseparable. The expanding global economy is stretching the carrying capacity of Earth, and we are tasked with learning again how to live a sustainable life. It is the most important project of our lives. But how do we achieve the necessary clarity to know the job that lies ahead? ❖

– Lesson Two: The Secret to Clear Water –

The secret to clear water in a goldfish pond is a matter of beauty and balance. This should come as no surprise. We learned from the Balinese that water is crucial to maintaining equilibrium in life, and this not only introduces grace and loveliness to our surroundings but also helps to hold our world together, even the microcosmic world of a goldfish pond in a garden. When we put a pond in a garden we tend to make the world more attractive and composed, and maintaining its clear water plays an important role in a pond's health. If we can keep our pond water clear, its myriad living components—plants, frogs, and goldfish—will fall in line like an army of obedient little ecological soldiers. Seated beside it, we'll also have a much better chance of seeing what's in the pond. With clarity, too, comes a more sagacious mind, which is essential if we wish to finally look beyond the pond and see greater meaning in the world around us.

A pond is considered to be in balance when it is in a condition called *homeostasis*, a term derived from the Greek words *homeo* (similar) and *stasis* (standing still). It has somewhat different meanings when applied to a person or society, but in a pond sense, if we turn to Wikipedia for our definition, homeostasis is "a system that maintains a stable condition of properties like temperature or pH." In water chemistry, pH is a measure of acidity or alkalinity, which relates to "hard" or "soft" water in a goldfish pond, and this turns out to be an important factor in achieving water clarity in a goldfish pond.

A commonplace synonym for homeostasis is equilibrium. It often appears in discussions about natural systems, society, or the state of a person's emotional well-being. When applied to human beings, equilibrium is sometimes mistaken for equilibrioception, which refers to the balance of our motor senses and is what keeps us from falling down when we are standing up. When we observe using all of our senses—motor and otherwise—not only are we less inclined to topple over, but we are better able to find meaning in the world. This acumen, a kind of sixth sense, is also called common sense, or more obliquely, horse sense. We tend to discover our balance by it.

Our perceptivity, which is basic to how we see our place in the world, is physically attuned to the wavelengths of energy that flow through the world. Passing energy is detected as sound by our sense of hearing in the 16 to 20,000 hertz wavelength range and detected by our sense of sight in the 0.4 to 0.7 micron wavelength range. We process the received information using our brains and, perhaps, with our minds and spirits, and thus do we comprehend the world. In this way, the "horse sense" so often commented upon may not, in the end, be so different from the Daoist philosophies of energy we earlier discovered in Vietnam. Both are a matter of awareness. Unless our receiving faculties are physically impaired—for instance, with poor eyesight or hearing loss—we all sense the world in pretty much the same ways. Yet, we uniquely interpret or make sense of the information we receive. This has much to do with the degree to which one has an open, engaged, and inquiring mind, and this is why two people can reside in the same world but experience it dissimilarly, such that it seems as if they are in two different worlds. This leads us back to the Hindu concept of *maya* (illusion), which instructs us that it's not a one-size-fits-all world in which each of us perceives things in exactly the same way. When we see things differently, they are different. Some people might not even notice a goldfish pond in a garden much less find a world within it. Contrary to popular thought, then, wisdom for this reason becomes a relative thing.

Of course, water clarity is important if one wishes to watch the goldfish swimming in a pond. A person needs to be able to see through the water, not just look at it. It should be that clear. Murky green water is the bane of many garden ponds and results from millions of tiny algae spores inhabiting the water and cluttering it with their accumulated biomass. The pond's algae make their appearance when nutrients, salts, sunlight, and carbon dioxide come together in just the right quantities for its growth. The right mix for algae, though, is not the correct balance for clear water in a pond. The color of green and its wondrous variances—verdant, moss, mantle, myrtle, fern, asparagus, pine, sap, and laurel—all beautifully describe the gardens and woodlands surrounding my goldfish pond, but they shouldn't describe the water in it.

If a pond turns green, we can easily replace its water with clear tap water. This is commonly done, but is a short-sighted solution to water problems. The pond will simply turn green again when the algae rebound, and then we've wasted a precious resource. Our task is to make sure that the balance of the pond is maintained so that its water won't turn green in the first place. A secret to keeping a pond filled with clear water is to starve the algae of what it needs to live, thus keeping it in check, and the best way to do this is by growing plants in the pond.

This is strange. Algae are plants, aren't they? Well, no. Like plants, algae produces chlorophyll (hence the green tincture), but it's not a plant. Nor is it an animal. Biologists have put algae into a unique kingdom called Protista, which falls outside the well-known plant and animal kingdoms. The term Protista was coined in 1866 by Ernst Haeckel, who, as we noted earlier, also came up with ecology. He employed Protista to encompass animal-like protozoa, plant-like protophyta, and fungus-like slime and mold, all of which do not quite fit the parameters of being either a plant or an animal; in other words, Protista is a kingdom of misfits. They survive almost anywhere there is water. Besides living in a person's goldfish pond, algae may be found in fresh or sea water, soil, trees, boiling hot springs, and deep down in polar ice. Scientists are looking for it on Mars. In addition to being prolific, algae are tenacious, and removing it from your pond is not a simple matter. It's much better to keep algae from proliferating in the first place, which is true for most problems in life, not just those of the Protista variety. A pond filled with green-colored water is another example of a world out-of-kilter.

At a global scale, algae, as small as they are, are hugely important. They produce more oxygen than all terrestrial plants in the world combined, even more than the Amazon rainforest. They are a food source for ocean life ranging from shrimp to whales. They capture and store vast amounts of carbon dioxide and thus are critical to mitigating energy imbalances in the atmosphere and the consequent warming trends being recorded on Earth's surface. Algae truly are a global player in climate change. But as important as algae may be for the food and energy balance of the planet, they have a limited and specific role to play in a goldfish pond and will show up in large numbers only when things have gotten out of balance.

When algae appear in the pond in such quantity that they turn the water green, we may be experiencing something akin to eutrophication, which is a concentrated bloom of algae caused by excessive amounts of organic or chemical fertilizers entering the water. This can easily get out of hand. Agricultural fertilizers and petro-chemicals used in America's industrial heartland enter the tributaries of the Mississippi River in huge quantities and eventually flow into the Gulf of Mexico, where they've produced a "dead zone" of hypoxic water covering eight thousand square miles of ocean along the coastlines of Texas, Louisiana, and Mississippi. It's an ecological disaster. A pond is unlikely to experience such a severe problem of eutrophication. A nutrient load at that level would not simply turn the water green with algae, but threaten to transform it into a slimy mass of furry tentacles that would throb in the water in a truly nightmarish way. A green monster might emerge from the pond and slither through the garden, leaving behind a sticky, wet trail of tendrils, shiny goldfish flopping on their tails and gasping for breath, and broken lily blossom petals. More likely, the pond simply is getting too much sunlight, which promotes modest algae growth but not full-blown eutrophication. Even so, algae at the lower concentrations of a green-water pond signify a problem of ecological imbalance.

An algae bloom works like this: (1) nutrient build-up from fertilizers or sewage; (2) rapid algae growth, also correctly known as "algae bloom," but I hesitate to use that term because it suggests something nice like a flower; (3) green water so thick that a person can almost walk across it; (4) sunlight not making it to the bottom of the pond; (5) underwater plants starve and die due to lack of sunlight needed for photosynthesis (6) the oxygen level in water is reduced because of diminished levels of photosynthesis, the state of hypoxia; (7) bacteria decompose the dead plant tissue and, in the process, use up the remaining available oxygen; (8) fish die from the overall net loss of oxygen in the water.

Immediate intervention is necessary in the event a person notices this kind of progression taking place in the pond. It happened once in my goldfish pond, and I rue the day I never saw it coming. A primary purpose of the pond, after all, is to provide a safe haven for goldfish, and we don't want to asphyxiate them because we haven't learned what we need to know about the importance of maintaining a proper balance for clear water. If the algae problem hasn't progressed all the way through the above series of steps, we can easily nip it in the bud by restricting the algae's share of radiant energy entering the pond.

There are many kinds of algae inhabiting our world. Some are parasites and feed off animals. Others are symbiotic and live on rocks along with fungi and lichens. The kind of algae one likely encounters in a goldfish pond are called autotrophic algae. They use the sun's energy to produce food through photosynthesis. Imagine if we could just walk outside and absorb energy directly from the sunrays and not have to eat food in order to stay alive? Think how easy life would be: no industrial agriculture, no feedlots in southwestern Kansas, no cooking or washing dishes or putting them in the dishwasher, no need for controversial genetically modified crops, no migrant farmers living in cramped trailers on Red Lick, no cravings for a midnight snack (maybe just a cup of coffee in the morning by a goldfish pond). But we can't live that way, nor would we want to. Only plants, algae, and cyanobacteria convert sunrays directly into biological tissue. In the process, they capture a total of about one hundred thirty terawatts of solar energy across the surface of the planet. That is more than six times the total amount of power annually consumed by the entire human race!

Since a pond's algae thrives because it consumes sunlight, one way to starve algae is to reduce its intake of solar energy by putting it on an energy diet. This requires blocking a portion of sunlight from entering the water and making sure the algae already in the pond don't absorb all the radiant energy that does reach it. The best way to accomplish this is by introducing competing plants into the pond. Submerged plants take up carbon dioxide in the water, oxygenate it, and out-flank the algae for the incoming solar energy. Floating plants, meanwhile, block sunlight from ever reaching the depths of a pond, thus diverting energy right away from the underwater algae. The right number and combination of aquatic plants necessary to maintain clear water is a matter of pond size, number of fish, depth of water, sunlight, and other pond-specific factors. It also is a matter of experience. There's no good substitute for carefully watching your pond over time to see what happens in it, just as there is no alternative to closely observing the world around us in order to see what is happening to it.

It's really quite remarkable how quickly a pond can change in color. One day I'm looking at a pea-green soup of algae in my pond, and before long it's as clear as a bubbling mountain brook. When we fill a pond with water for the first time, algae will almost certainly thrive in it. They were among the first wave of settlers in my goldfish pond, right after the pioneer frogs. It takes a while for the plants to get established, kick into gear, and do their balancing act. But ecological balance in a pond is a work-in-progress and not a beach ball poised for a moment on the nose of an otter; it's an ongoing process filled with starts and stops. Ecological balance is a dynamic state. We might find ourselves frustrated at first by a delay in water transparency and be tempted to add clear water from the garden tap to speed things along, but it's not a desirable thing to do. It will only make things worse and stretch out the diaphanous timetable. Patience is required in this case. A few weeks after filling it with water, though, and a properly outfitted pond, with the right mix of plants, should be clear as a crystal ball. When this happens, the pond is oligotrophic, which means the water is translucent, and the frogs, fish, plants, and someone sitting on a nearby bench are all beneficiaries.

Despite our best efforts, a pond will probably turn green again with the start of each new growing season, just as it does the first time we fill it with water. Green is the color of spring, heralding warmer temperatures, the world coming alive again, new buds on trees in the forest, foliage sprouting in the garden, the planting season, warm sunshine, and the beginning of another cycle in the goldfish pond. Green is life. But green shouldn't actually describe the water color in the pond over the course of a year. Invisible should describe the color of water in the pond most of the time. But even invisibility shouldn't be taken too far. It, too, may suggest an environmental problem.

Some of the clearest, most invisible water I've encountered anywhere was in the lakes of the Adirondack Mountains in upstate New York. I was hiking in those ranges a number of years ago and camped by lakes that were so clear I could see right through them. Looking into them was like peering at a pane of glass. Try as I might, I couldn't see a single thing in the lakes apart from the rocks that lay on their empty bottoms. That was because nothing was alive in the Adirondack lakes. They were dead clear, like the Dead Sea, which is devoid of all plant and animal life. Unlike the Dead Sea, though, which suffers from toxic concentrations of salts and other minerals, the Adirondack lakes were suffering from low pH due to acid rain. They were out of balance because pure rain no longer commonly exists on the planet.

The scientists hadn't yet figured it all out, and I didn't realize at the time that the Adirondack lakes were clear because the water chemistry was wrong and they couldn't support life; they just looked clear to me. We don't want something like that happening in the goldfish pond. The acid lakes in the Adirondacks contained neither algae nor phytoplankton, and thus were missing a basic building block in a pond's food chain. In this case, it was no algae, no fish; but in other instances, too much algae and also no fish. Again, it's a matter of balance; as Bob Marley once wailed: *No woman, no cry.*

The types of aquatic plants we put in the pond have a great deal to do with its health. Plants, then, are one of the big secrets to clear water. A natural pond in the region where I live has a wide variety of native plants inhabiting and filtering it. These might include bulrush, cattails, bassweed, floating heart, watercress, arrowhead, crow's foot, hornwort, golden club tail, or mosquito fern. Many of these species are too large, though, for a garden pond, and would look out of place in it, unless we have an especially large goldfish pond. A person also might want to include in the water some showy, non-native aquatic flora with bright flowers and interesting foliage. A local water garden store will have a good selection of both native and exotic plants to consider for the pond. Keep in mind, though, that if we bring an alien plant species into the garden, even a lovely aquatic one, it might one day jump out of the pond and head straight for the surrounding forest, invading it and overtaking the native flora like Odysseus' soldiers spilling out of the belly of a Trojan Horse after crossing the Aegean Sea, then sneaking into Troy and conquering all of Turkey.

I've introduced some Japanese irises at the edges of my pond and they form a mass of deep purple that spills into the garden. Irises are a good marginal bog plant for any pond and make a smooth transition into the surrounding landscape. I've also submerged water lilies in pots in the pond. Their floating pads cover much of the water's surface, reducing the amount of sunlight entering it, and their white blossoms sparkle in the morning light, brightening my world whenever I have the occasion to sit by them. The lily pads also provide great platforms for frogs to rest upon and from which to launch out at passing insects. A water lily is a multi-tasking pond plant. Otherwise, I have a few individual specimens of select aquatic species, which I've placed here and there in the water to provide accent. The lotus, though, is my favorite plant in the pond, a tall creature with showy pink blossoms held upright on skinny, prickly stalks. It's the queen of my goldfish pond.

Check with the local pond store about the right mix of aquatic plants for the pond; there are too many of them to mention here. What works best, though, in my experience is a nice blend of floating and underwater plants, marginal plants and those that will occupy the middle reaches of the pond, as well as flowering plants and others with distinctive foliage. Mix it up a bit. A little water lettuce floating here and there in the pond adds a nice touch, like sprigs of basil flavoring a bowl of Tom Yum soup. By no means, though, should anyone introduce duck weed or fairy moss into the pond. Those tiny plants are cute and harmless-looking, with innocent-sounding names, but they reproduce like mad and will become invasive to the point of despair when it comes time to eradicate them.

So far in this lesson, we've introduced terms like homeostasis, equilibrium, eutrophication, and hypoxia to our pond lexicon. They all are part of the secret to pond health and clear water. We have to also consider nitrogen and ammonia cycles. Assuming the pond is more or less a closed system—for instance, water doesn't enter it from the outside world unless it rains or we put it in—and we don't add a lot of nitrogen to it in the form of chemical fertilizers, the nitrogen and ammonia in the pond naturally circulates among the bacteria, plants, and fish in the water. When a fish eats a plant or a food pellet, its digestive system breaks down the proteins into ammonia, which is excreted from its gills as well as its urine and feces. Ammonia is also released from organisms that die and decompose in a pond. We'd notice this right away if a dead raccoon or deer were floating in the pond, but it more likely comes from deceased fish or frogs that we might not see because they tend to disappear beneath rocks or plants. A pond creature, like most animals, will burrow in some place when it's time to die. Gases eventually build up within the carrion, causing it to rise to the surface, but we'd probably smell a rotting corpse before spotting it floating on the pond's surface. A naturalized pond, like the rest of life, is filled with both beauty and tragedy. In any event, and with or without anyone noticing, the decaying tissue of a dead creature is attacked by heterotrophic bacteria and converted into toxic ammonia.

Regardless of its source, too much ammonia is deadly to fish. Fortunately, the nitrifying bacteria naturally present in a pond work to convert the ammonia into nitrites, and the aerobic bacteria then turn the nitrites into benign nitrates. Plants absorb these nitrates as additional food, and the fish, of course, eat the plants. And so it goes. The nitrogen cycle in a pond is a bit like reincarnation in India. It's a timeworn refrain: what goes around comes around. It should be noted, then, that yet another secret to balance and clear water in a goldfish pond is the presence of beneficial bacteria. And this is where a pond's filtration system comes in. A lot of people think that all one needs is a filter and—abracadabra—clear water! Water clarity certainly has its little mysteries and secrets, but it's not magic.

If we have a healthy stock of plants in the pond, we have a natural filtration system already working for us. The stems of the submerged plants provide a clinging surface for algae, holding them captive so they don't float freely in the water and thus help keep the pond clear of their cluttering green color. The marginal plants, meanwhile, block the creep of dirt, nutrient-laden mulch, and other detritus from entering the pond. The floating and submerged plants compete with algae for sunlight. These things are a huge help. Still, we may need additional filtration for a large pond, or even a small one that we want to keep consistently full of clear water and goldfish. Mechanical filters are one available type; they trap and remove solids and waste, and are fine for some ponds, but they tend to clog with prolonged use and some of them require chemicals like zeolite to remove impurities from the water. At a commercial scale, mechanical filters are used in the big ponds found at golf courses and hotels, where you also find fairway hazards and good luck coin-tossing pools. If a person decides to go with a mechanical filter for a garden pond, fine, but keep it simple. There's no need to throw into the water a glob of additional clarifiers and flocculants like they do at the golf courses, which will turn the pond an artificial blue color that some people enjoy, but which has nothing to do with the reflection of the sky in the water on a bright summer day.

An ecologically better way to keep the pond water clear is to use a biological filter, which works by harnessing naturally-occurring bacteria in the water and letting them do the job of filtering it. The biological filters come in all sizes, but they employ the same basic design: a pump to draw pond water into a filter unit, an outer chamber filled with a sponge-like material that traps sediments sucked into the filter, and a series of inner chambers filled with progressively finer grades of porous rock or gravel, which act as hosts for the bacteria. The sky's the limit in terms of the kinds of biological filters on the market, with some real Cadillac models for sale, but the principle is always the same: we set up a natural bacterial colony to purify the water in the pond and keep it clear.

If, despite all best efforts, the pond lapses into a pea-green soupy mess, fear not. When we again achieve the correct balance of energy, nutrients, and plant and animal organisms, the pond, like other natural ecosystems, will sort itself out. All pond systems have cycles, and the water will occasionally get murky. It's nothing to worry about. If worst comes to worst, though, and one has to remove algae from the pond in huge quantities, it's nice to know that at least we can do something with it. For one, we might eat it. Algae contain high levels of protein and vitamins. I've never eaten my pond's algae, and so I can't attest to its flavor, but if it's anything like the oceanic seaweed called *nori*, which the Japanese use to wrap sushi or put into miso soup, it should taste okay. We can use algae to thicken shampoo, ice cream, soup, or puddings. It reputedly makes a good paint or dye. Algae are believed to cure illnesses like gout, diarrhea, and maybe even some forms of cancer. We could brush our teeth with it. Algae is being processed into jet fuel and diesel and, according to some estimates, may produce up to one billion gallons of fuel within the next decade. Of course, if a pond is producing so much algae that a person is thinking about manufacturing biofuel, they probably have a much bigger problem on their hands than clear water. ❖

CHAPTER NINE

I'm not a serious bird-watcher, not in the sense of having a list to compile or anything like that, but I like to watch birds as much as anyone. I especially like it on sunny afternoons when yellow finches, indigo buntings, and bright, red cardinals cavort in the sky. It's like a rainbow has fractured and little broken pieces are floating down into the garden.

Goldfish are Wildlife?

I've discovered over the years that the goldfish in my pond will do fine without my feeding them. This is good news because I'm often absent from home. The algae, insects, and micro-organisms in the water keep them fed when I'm away. Still, I like to watch them eat, and George and I often include goldfish-feeding in the routine of our morning coffee at the pond. Most people who maintain ponds enjoy feeding the fish they put into them; this is also true for industrial-scale aquaculture ponds meant to raise catfish for a commercial food supply, livestock ponds, or pay-lakes stocked with bass and bream for recreational inch-by-the-dollar fishing. My brother-in-law trained the bluegill in his farm pond to eat right out of his hand, and now I'm waiting for him to show me how they will do little tricks in the water. Meanwhile, feeding the Koi in a Japanese garden pond is the pinnacle of good luck, and we always need more of that.

My fish feeding follows a more or less regular routine. I bang a piece of bamboo gently against a rock in the water to attract the attention of the fish and let them know that it's mealtime; in other words, I ring their breakfast bell. Like Pavlov's dog, they come when they hear the sound of food. It's a colorful sight. The fish leave their hiding spots, swim to the edge of the pond closest to the bench, and school into a swirl of writhing color, all lunging and jostling for position as if they were speedo-clad competitors at the starting line of the Honolulu Iron Man triathlon. The fish in my pond are frenzied at the prospect of victuals. Their shapes blur and the colors bleed into one another as they furiously compete for the best seat at the pond table. Of course, there is no *best* spot when I feed the fish. I throw food pellets wherever I want in the pond, and they simply swim toward them. It's a slight power I hold over the goldfish. They won't fetch your slippers or chase a stick like a dog. Fish won't jump into a person's lap when we call them or purr and arch their backs when they're petted (not that I've tried to pet a fish). They can't talk like a parrot or sing like a canary. And they won't do tricks like a little pot-bellied pig. You can't *ride* a fish — not even a seahorse — or train a school of them to pull a plough through the garden. Fish are not like other household animals. They do whatever they want to do, until feeding time, when I have a little control over them.

At first, I throw a few pellets into the water to see which of the fish are most on the ball. The plucky daredevils come forward first and grab these initial tidbits of food. Then I throw in larger handfuls of pellets. This gets the fish school really thrashing about, and the water roils with their impatience. Despite the apparent disorder, there is logic of sorts to my pond's dining queue. The largest fish grab the first pellets and sink into the water to digest them. The fish next-in-size then swim forward and grab some food. They all form perfect little "o" shapes with their greedy mouths and vacuum in a pellet floating on the water with the efficiency of a pond skimmer. After ingesting their meal, the mid-size fish also drop into the water to dispatch it. Then the smaller fish tentatively come forward. And so on down the line of fish until the very smallest ones, waiting like timid students in a reformatory school, get their chance to eat at the pond cafeteria. By this time, though, the larger fish are hungry again and ready for more food; they'll push and shove their way again to the front of the food line. *They jump queue*. If it were entirely up to the larger fish, the smallest ones would never get a chance to eat. The big fish are school bullies that way. But I'm the one in charge, and like a public school principal keeping students in line, I always make sure that some pellets land close to the shy and small fish. If the fish are very tiny, their mouths can't take in a full-size pellet and they're forced to nibble away at the edges of it. This, of course, gives a larger fish the chance to steal it from them. It happens all the time. Mealtime is an unruly period in my pond's school of fish.

Because I have a slight influence over the fish at feeding times, I also make them do things they might not otherwise think to do for themselves. I might distribute the pellets across the water to move the fish in a precise line, as if they were dancers in a piscatorial ballet. I can choreograph a specific fish or the entire school of them in this way. The rippling avenues of fish chasing food pellets across the pond renders a colorful impression on its surface much like the brush strokes of Monet in one of his *Nymphéas* paintings. I sometimes keep a music player nearby and play a "Bach for Breakfast" CD during one of these morning performances. You can imagine what that is like.

Punctuating the aqueous canvas of my goldfish pond are the water lilies. One of the loveliest sights in my garden—possibly in the entire world—occurs when a goldfish swims through the reflection of a lily blossom so that I see them superimposed for a moment as a single thing. It makes me think of Alice stepping into the Wonderland looking glass: *Will the fish be the same when it emerges on the other side of the efflorescence?*

When I feed the goldfish in my Wolf Gap garden pond, I follow a family tradition of fish feeding. I grew up in Michigan, and my parents would often stop at trout hatcheries on our vacations to the northern parts of the state. The Fish and Wildlife Department maintains the hatcheries to stock the forest streams with rainbow and brown trout for recreational fishermen. Arriving at the hatcheries, we kids would run off to the fish food dispensers that looked to us like bubble gum machines, insert a quarter, and grab the handful of pellets, and then race to feed the trout. There were hundreds of fish in a stocking tank and they exploded with activity when we threw in some food. That many fish all scrambling for a handful of food makes for quite a mob scene. I still enjoy trout feeding whenever I'm at a hatchery; it's always exciting, but as a general rule I feed fish whenever I'm out fishing—purposely in order to attract them, for example when chumming in the ocean with scraps of bloody fish flesh or shrimp, but mainly because fish I'm after steal the bait before I notice and swim off, leaving me with an empty hook and dashed hopes. I'm not a very good fisherman for this reason. Still, feeding fish for the fun of it and with no intention of catching one is something I commonly enjoy doing at my goldfish pond in Wolf Gap Holler.

It's not just that I feed the goldfish in my pond, though, because by keeping them captive, I also feed them to raccoons that prowl the gardens at night, and, I suspect, to my cat Olive, although I haven't yet caught her in the act of eating one. The raccoons don't seem to bother the fish when they're young and tiny. I guess it's not worth the risk of getting wet to go after the minnow-size fry. I've noticed that once a fish attains a decent size, say about a hand-length, its days are numbered. They tend to mysteriously disappear overnight. One day they're happily swimming about the pond, minding their own fishy business, and the next day they're gone. Occasionally, in the morning, I find a gnawed-upon fish that is floating belly-up in the water, and that is always a sad sight. It doesn't happen often, fortunately, and I don't want to give the wrong impression, but when it comes to feeding the fish in my pond, it can go either way.

A person living in the woods is surrounded by wild animals. Many of them make their way into our gardens and over to my goldfish pond. A family of deer, for example, resides in the nearby forest and grazes at night in our gardens; they eat roses and lily buds just prior to blooming, and new plant foliage anytime, but mainly they like to consume the vegetables ripening in Jennifer's kitchen garden. The deer treat our garden like a salad and afterward go to the pond for a quiet after-dinner drink. And they are just the vanguard of a nocturnal zoological invasion. Ground hogs, raccoon, skunks, and rabbits also frequent our place as uninvited guests in the garden. They know they're not welcome and only sneak in at night when we're asleep. I sometimes wonder why we don't get a dog to watch over the place; it seems like a good idea, and Jennifer frequently brings it up, but the whole idea is complicated by the fact that we travel a lot. Anyway, I enjoy seeing the deer in the forest, even when I know they sometimes make themselves at home in our gardens at night.

It could be much worse. I spend a lot of time in Nepal, where villagers suffer the nightly visits of elephants, rhinos, tigers, and large troops of bellicose macaque monkeys. It's easy to imagine what kind of damage *they* can do in a garden. The big animals sometimes run right through a village, crushing homes and trampling people sleeping in their beds, while an army of simians running amok in a cultivated field can devour an entire harvest of food in a single night, putting people on the brink of starvation. Clearly, my wildlife problem in Wolf Gap Holler is nothing compared to all that. It's frustrating, though, to wake up in the morning and see our gardens torn up by the animals, but that's the price a person must be willing to pay when they build a home in the middle of a forest, whether it is in a small Kentucky valley or among the villages of the Himalayan foothills.

People in Nepal began leaving their highland homes in large numbers in the late 1960s to settle in the low-elevation forests near the Indian border. The human population by then had exploded in the hills, environmental conditions worsened, and the small hill farms were in serious decline. Nepalese officials promoted the lowland border region, known as the Tarai, as a new frontier for agriculture and encouraged people from the mountains to move there. Health officers sprayed DDT in the Tarai to eradicate the malaria-carrying mosquitoes that were endemic in the jungle. Prior to the spraying campaign, only native Tharu people could live in the Tarai because they had built up a genetic resistance to the disease over generations. Once the forests were cleared of mosquitoes, though, the highlanders came down in droves. They cleared tracts of forest, built villages, cut and burned trees, planted pasture, cultivated fields, bred and grazed livestock, and provoked a deadly confrontation with local wildlife. The health workers initially had been concerned about mosquitoes and malaria, but upon arrival the settlers found that they faced graver threats: deadly vipers, troops of thieving monkeys, wild elephants, tigers, and the knock-about rampages of near-sighted rhinoceros. The newcomers knew little about these wild animals, having grown up in an alpine world, and were rightly afraid of them.

Much of the megafauna in Nepal is rare and in need of protection. Especially threatened are the big Tarai animals: Royal Bengal Tigers, Indian One-horned Rhinoceros, Mugger Crocodiles, Golden Jackals, Clouded Leopard, and Wild Buffalo. Changes have taken place among their habitats, and some populations are now so depleted that these species are close to extinction. The large animals are perceived as threats by villagers living in close proximity to them. Some people began killing them out of a sense of survival. Others poached them for commercial profit. The natural range of the wildlife declined as more forest gave way to farms, and this, too, hastened their demise. In 1973, faced with a bleak outlook for its rapidly dwindling numbers of endangered wildlife and under pressure from conservation groups, His Majesty's Government of Nepal established Chitwan National Park on 550 square kilometers of jungle. It was the country's first sanctuary dedicated to protecting animals and their natural habitats.

The native Tharu living in the territory were forced to vacate their ancestral homes and establish new settlements across the Rapti River. Highland immigrants, meanwhile, continued to descend on the forest. The problems that people faced with wildlife didn't go away with the new park; animals still crossed its boundary in pursuit of forage or mates, and villagers continued to enter the reserve to gather thatch and livestock fodder or on illegal hunting expeditions. The perimeter wasn't fenced, and although it existed on government charts, the boundary line had no place in the mental maps of local people, much less the instinctual behaviors of resident wildlife.

As people piled onto the park perimeter and animal habitat located outside the reserve further diminished, the human and wildlife confrontations intensified in the Tarai. On busy nights, dozens of elephants or rhinos might barge into a village while its inhabitants were asleep in their flimsy thatched homes. I can only imagine what that must have been like for them. Elsewhere, monkey troops descended onto farmers' fields and devoured everything in sight, putting to rest the notion of a harmless primate.

The deer foraging in our Wolf Gap garden, picking delicate sprouts of lettuce or drinking politely at my goldfish pond, provide little insight into these life and death matters in Nepal. In a typical year, tigers might kill five or six villagers, rhinos a few less than that, and the casualties from rampaging herds of elephants, while less frequent, are reported in some years to be in the double digits. Despite such events, the Nepalese villagers actually have less to fear from the wildlife than they do of them. Poachers regularly trespass into the park to kill animals for profit, inspired by a global demand for animal parts. A kilogram of powdered rhino horn, for instance, may fetch up to 60,000 USD in South Korea. Such illegal activity draws Nepal's wildlife and human residents into a ruthless international trade in endangered species that annually is estimated to be worth billions of dollars and globally involves the destruction of millions of plant and animal specimens. There are no winners in this violent contest between people and nature. It serves as yet another lesson in knowing the importance of boundaries in our world.

One evening in Nepal I encountered a stealth elephant. I was walking on a riverside path near Chitwan National Park on my way to a lodge where I was to spend the night, and the jungle was filling with twilight sounds: cicadas, tree frogs, and shrill night birds, and also the occasional cough or grunt of a large animal. I sensed a presence behind me and heard a soft "cluck-cluck" sound as I walked. When the noise persisted I turned around to find myself squared off with the trunk of a huge pachyderm. I had to crane my neck to peer into its eyes. It was like looking at the side of a barn. The proboscis was the size of an oak tree. Each ear was bigger than a table top. The feet were like butcher blocks. What surprised me most about the elephant, though, was not its enormity but how quietly the animal had come up behind me. I never heard its approach. I had only vaguely sensed its looming presence and then heard the "cluck-cluck" warning sounds made by the mahout, who was atop the beast and notifying me of its presence on the path. I moved aside and let the elephant go by. It was like a ship passing in the night.

Fortunately, the elephant wasn't a wild one or things might have turned out differently for me. I was on the outskirts of a popular tourist village where elephants were kept to ferry visitors into the park on sight-seeing expeditions, and I didn't have any reason to worry about being attacked by a wild one. Still, it's common enough for people in Nepal to get trampled by elephants near the park, even those living in the populated villages. That's a price some people pay when they put their home and garden in the middle of a forest. It gave me pause. What, indeed, would I do if such a huge and silent animal were to walk through my garden in Wolf Gap Holler, knock over trees, and plunge into my goldfish pond? It's a question I've had to ask myself in Nepal, but which, fortunately, never comes up in Kentucky. However, I live within the natural range of the American black bear (*Ursus americanus*), and recently, I saw one crossing Red Lick Road early one morning while on my way to work, so it may be something I will need to think more about in the near future.

In many ways, my goldfish pond is the epicenter of wildlife action in Wolf Gap Holler. It's another kind of natural entertainment on offer in my garden. Most of the activity involves fish or frogs and flowering water plants, but one afternoon I came home to find a Great Blue Heron wading in the pond. It was a pterodactyl moment. When it heard my truck pulling into the gravel lane, the great bird took flight and the sight of its mighty wings lifting from the tiny pond was epic and prehistoric. Another time, I was sitting on a bench partially obscured by a magnolia tree and watched as a hawk circled above the pond. It must have seen the fish in it. The flight pattern of the bird got tighter in the sky as it descended lower toward the pond. I crossed my legs as I watched the spiraling descent of the hawk, and my slight movement was all that was required for the bird to pull out of its orbit and fly away. Hawks have such amazing powers of sight. They can see a rabbit a mile away. Twice we've had hawks fly into our screened porch, ripping a hole in the screen each time and getting caught inside like a giant bird inside an aviary. The first time it was a Coopers Hawk, which is a pretty impressive bird, and it took two days for it to finally realize that it could simply walk out the open door and fly off into the woods.

The second time a hawk got caught in our screened porch was

when George was house-sitting for us while we were away in Asia. About a month into our trip I got an email from him, which is not the same as having a conversation with George by the goldfish pond, but at the time I was on the other side of the world where emails come in handy. In his message, George related how he had gone out to the porch one day to find a hawk inside it. *Two* hawks flying into the *same* screened porch? What are the odds of that? It's hard to believe. We must have built our home in the middle of a busy hawk flyway. In any event, George found it was easy to extract the hawk from the porch that time. It was perched on the rail of a chair, and when he showed up it moved onto the hammock, where it swung back and forth for a while. Leaving the hammock, the hawk hopped up to a ceiling fan and landed on one of its blades. The momentum of flight caused the fan to go round and round. The hawk apparently got dizzy riding the spinning fan blades because it toppled conveniently into an empty planter resting on the floor. George put a cardboard lid on the planter, scooted it outside, and set the hawk free. He wrote to me that when the bird finally realized it was no longer captive, it ruffled its shoulder feathers, like the Coopers Hawk did, as if to shrug off a bad memory, and then flew straight away into the woods without a backward glance.

George was so moved by the hawk episode that he wrote a poem about it, which he presented to me upon my return to Wolf Gap later that year. Here's an excerpt:

Screen Test (2011)

A twister, a ripper,
Speedy and compact,
A screen-piercing missile
You are not a houseguest
Who understands

Entrances and exits. What have we crashed through
What desires have we
Impaled upon
What traps entered
Never to leave?
What have we
Followed recklessly
With speed joy resolve
Only to find
Ourselves confused.
Uncomforted by any past
Frightened of any future
Ignorant of how to soar?
What falconer will
Show us the way?

Apart from the occasional hawk or heron, the birds flitting around my goldfish pond are a welcome sight. They come in from the woods in great numbers: cardinals, yellow finches, indigo buntings, grosbeaks, cooing doves, tiny nuthatches and chickadees, flickers, noisy junkos, kingbirds, and mockingbirds. About 375 species of birds are native to Kentucky or pass through the area, and we get a good number of them in Wolf Gap Holler. One of my favorite species is the pileated woodpecker, which stands almost twenty inches high and is downright cartoonish with its bright red crown and Woody Woodpecker looks. When I see one in a forest tree, I can't help but hear Woody's sly and rather malicious-sounding laughter. The *knock-knock-knock* of a pileated woodpecker pounding on a hollow log can be heard throughout the valley. But the big birds do a tremendous amount of damage when they go after the carpenter bees that inhabit the wood of our homes, removing large splinters of it with their strong beaks. I prefer that the woodpeckers remain in the forest, where they belong, and stick to hunting insects in the dead trees. I'm not a serious bird-watcher, not in the sense of having a list to compile or anything like that, but I like to watch birds as much as anyone. I especially like it on sunny afternoons when yellow finches, indigo buntings, and bright, red cardinals cavort in the sky. It's like a rainbow has fractured and little broken pieces are floating down into the garden.

George is much better at identifying birds than me. He's been feeding them on his property for so long now that they recognize the sound of his truck pulling into the lane and fly to the cedar posts near his house where he places their food. I used to feed the birds in my garden along with the goldfish in my pond, but the raccoons managed to get at the seed in the bird feeders, just like they do the fish in the water, and so I halted the practice. The raccoons have plenty of time at night to figure out these kinds of things, and it didn't seem to matter how cleverly I rigged a bird-feeder, they always found a way into it. The squirrels soon joined them. It got to the point where I just stopped feeding the birds. And then, of course, there are our cats. With them lurking around, we don't want to encourage the birds to make a home in our garden. It's okay if a bird swoops in for a quick meal and then leaves, but we prefer that it not live in our garden for its own good. It's too bad, though, to put restrictions on birds like that. Fortunately, I can still enjoy birds living in the forest when they fly through my gardens. It's always a colorful sight, and, of course, I can hear them in the garden whenever they sing in the woods. For the time being, then, I've restricted my wildlife feeding to the goldfish in the pond. But they aren't really wildlife, are they? ❖

CHAPTER TEN

At minimum, heating my home with a woodstove gives me a chance to turn on its head the old saw, "it's hard to see a forest through its trees." The popular phrase suggests that it's difficult to discern an overall pattern in the world because we are distracted by its details. In fact, it is because I take an occasional tree from the forest around my goldfish pond and burn it in my woodstove that I become more aware of how I fit into the overall scheme of life on Earth.

Living with a Woodstove

The Wolf Gap forest provides valuable natural resources and hosts a rich diversity of native plant and animal life. Wild game, mushrooms, ginseng, and other foods come from it. We walk among its trees for recreation and quiet reflection. I transfer ferns, dogwood, and hemlock trees from the forest to my gardens. It also provides me with firewood that I burn during the winter months.

Most folks living in Wolf Gap Holler heat their homes with wood. When a person uses a woodstove, they *live* with wood and are committed to a life with trees. This connects us not only to the incandescent power of a forest but to one another; to the turn of seasons and all the necessary preparations for winter; to a community of land-owners, loggers, and lumbermen; to friends who can help out when it comes time to cut, split, and stack firewood; to sawmills, foresters, chimney sweeps, and stone masons; to the hardware store that sells chain saws and oil; and to pickup trucks and tools such as splitters, mauls, and axes. Some people believe that by heating their homes with wood they free themselves from a dependency on other people, but that just isn't true. It's also mistaken to think that a woodstove will solve global energy problems.

Over the years, I've found a main reward of heating a house with a woodstove comes from sitting in front of a cozy fire on a winter night, and knowing the heat that radiates into the room comes from a tree that once grew nearby. It's both an aesthetic pleasure and a direct affirmation of our dependency upon Earth. I don't think one could have such a realization with central heat, which pours out from everywhere at once so that it seems to come from nowhere at all. When a person heats their home with oil, gas, coal, or electricity, a more obtuse arrangement connects them to the source of their energy. It would be difficult for such a person to follow a wavelength of energy leading from the sun to the interior of a home when it must pass through fossil fuels, which span geological time; run a gauntlet of refineries, transmission lines, highways, and pipelines; and navigate an opaque network of global finances before finally reaching their home. How could one possibly keep track of such a circuitous path of energy?

Amid overwhelming obfuscation, most of us choose not to think at all about what it really takes to heat our homes. We just flip a switch and pay the utility bills. When we heat our home with fossil fuels, we burn the carbonized vessels of solar energy: coal seams in the Appalachia Mountains, petroleum pooled in the Arabian Desert, natural gas in shale deposits beneath the city of Fort Worth, Texas, tar sands in Canada. These vessels require millions of years to form, and this is what makes them non-renewable; their geological formation extends far beyond human time. The formation of fossil fuels is so distant, in fact, from the timeline of my own life that it is hard for me even to imagine it. With a woodstove, I am burning a vessel of solar energy—a tree—but it replenishes at a rate closer to my own time. I can watch a tree grow. And I can sit in its shade. It provides nesting room for songbirds until I might cut it for firewood.

Living with a woodstove engages a natural synchronicity. Hardwood species, such as the white oak, ash, or hickory that are native to Wolf Gap Holler, make excellent firewood. They also require about forty growing years to get big enough to produce a decent amount of firewood, at least enough to justify cutting one down for the sake of its combustible fuel content. That is a much shorter turnaround time than petroleum or coal, but still it stretches the energy equation across much of a person's lifetime. People who think they are using renewable energy when they heat with a woodstove are right, but it's not as simple as that. Generally speaking, most people don't have sufficient forest land to sustain their long-term firewood needs in any reasonable fashion. It takes several dozen acres of mature hardwood forest per woodstove for it to work out properly over the life of a household. Owning less than that, I would need to cut most of the trees on my property and manage it as a wood energy farm, but then I'd be living in a place akin to a logger's camp and not among trees that conform to natural laws. A cord of firewood requires about fifteen ten-inch diameter trees. It takes several decades for a tree to reach that size. It's common for me to burn more than a cord of wood in a typical winter, and I live in a relatively warm place, so where would all the firewood come from if we imagine it supporting a significant segment of human society living in cold climes?

Like other woodstove users, I'm always scavenging around for firewood. It's a year-round effort for me to procure a winter's supply of cordage. Every once in a while I might obtain a tree from my property when it suffers from old age or is blow down in a storm, but it's never enough. Not even close. It's certainly not a sustainable source of firewood. Sometimes a neighbor or friend lets me onto their land to cut firewood, but frankly, most people in Wolf Gap heat with a woodstove like I do, and so we're all in the same basic inflammatory relationship with trees, and hence, with one another because we're all competing for firewood. The likelihood of getting it from a neighbor's farm is slim. I approach the utility company trucks on roadways when their crews are out cutting trees along a transmission line and obtain some wood that way; the utility people are always looking for a place to dump the wood, but it tends to be a mixed lot of cuttings that aren't always good for a woodstove. Occasionally, someone I know in a nearby town will need a tree removed that's gotten too big for their yard or neighborhood. But these are infrequent encounters. And it's not just me out there scavenging firewood. There's an entire army of fagot collectors living on Red Lick and in Wolf Gap Holler.

My own scavenging behavior sometimes includes cleaning up after logging operations, which is a tricky subject in ethical terms. Many of the local lumbermen don't follow sustainable logging practices; they use bulldozers and front-end loaders to knock down and remove trees, generally clear-cutting a lot, because it's much faster than felling them with a chainsaw. But the machinery does an awful lot of damage to the soil in a forest. A commercial timber man is typically after the trunk when they go into the woods for a tree. The upper canopy of limbs and branches is left behind. Depending upon species, it provides excellent firewood. When I scavenge behind a logger's operation, though, I benefit from timber practices that lay waste to a forest; hence, my dilemma. I justify my actions by thinking the arboreal remnants would go to waste if I left them alone, but that's not true either because they would decompose and nourish the forest. To wit, this is another of life's many contradictions that I've had to come to terms with.

When the pickings are slim I buy wood from some local guys who cut and sell it for pocket money. A few are raising a family that way. The woodmen cut, split, and haul a truckload of firewood for sixty dollars. It doesn't seem like enough pay. The work takes much of a day with two men laboring on it. Sometimes I feel like I'm exploiting them just as I do the children in distant countries who stitch the clothes I purchase in a local store. I know the woodmen, though. They live near me, unlike the kids who toil anonymously in a distant sweatshop, and so they are closer to mind if not heart. Moreover, I've been buying firewood from some of the same guys for so many years now they consider me a regular customer and give discounts for multiple pickup loads. I pay cash so they don't need to report the earnings for tax purposes, but still, it doesn't seem right that they earn so little when the work is so hard.

These days I find I'm buying more firewood than ever before. With advancing age, it's become harder for me to keep up with all the work of it. Thoreau wrote that when you heat with wood it warms you twice: once when you split and haul it, and again when you burn it in your woodstove. That's true, but it doesn't make the work any easier. Thoreau went to Walden pond at the age of twenty-eight and wrote *Walden* as a young man. I no longer have the energy of my youth. Moreover, the Red Lick forests open to firewood collection are diminishing fast, succumbing to suburban encroachments and gentrified mini-farms. And with more people burning wood for heat, thinking perhaps they are helping the planet by reducing their carbon footprint or just because they enjoy sitting beside a woodstove on a cold day like I do, the scavenging rights for firewood are distributed across a wider segment of society. All of this makes it harder to come by. In any event, it's clear to me that more work will be required of me in the years ahead if I'm to continue to take care of my home heating needs with a woodstove.

However you measure it, firewood is not a sustainable source of energy for all of humanity. Not if it means cutting mature trees in large numbers, not even when the trees are dying in the forest or where an individual has enough timber land to meet their own firewood needs, and especially not when a majority of the world's population still relies upon it as their primary energy source. Life and land already suffer where household demand for fuel wood exceeds the capacity of the landscape to provide it. People living in the Himalaya, for example, spend more time with each passing year gathering a supply of firewood and must walk further into the forest every time they go out to collect it. On average, a villager in Nepal can expect to spend a couple hours a day combing the mountainsides for brush and tree limbs to burn in their hearths at home. In many places, wood gathering has contributed to degraded hillsides stripped of vegetation and marked by soil erosion and landslides, and this in turn leads to declining farm productivity and the out-migration of impoverished villagers to Nepal's towns, the southern jungles, or distant cities located in Southeast Asia or on the Arabian Gulf, where they strive for a better life. Forest damage from firewood gathering in my part of the world hasn't escalated to that point, but it might if most people actually were to heat with woodstoves instead of relying upon fossil fuels to heat their homes. It all begs the question of a new energy model for human society.

In the Sahel region of Africa, among the countries of Niger, Chad, and Mali, where the energy needs of the local populations surpass the carrying capacity of the landscape, firewood collection is linked to the global problem of desertification. Coupled with recurring drought and livestock grazing, the clearing of savanna for firewood creates a place devoid of greenery, filled instead with escalating human misery, war, and ecological impoverishment. The sandy edge of the Sahara Desert is drifting closer to the equator with each passing season. Still, more people in Africa are scouring the countryside every year for the dwindling wood supplies, competing with livestock and one another for the tiny fragments of biomass that remain on the desert's tattered margins. It's a matter of survival for most people living in the Sahel, amid a growing disparity that makes life more tenuous for everyone. A degraded landscape amid the intertwined tragedies of poverty, political unrest, and famine becomes an all-too-familiar refrain heard in many parts of the world.

This raises the troubling prospect of biomass as a sustainable energy resource for the 21st Century. It's in the news a lot. And it involves much more than firewood. Some industrial farmers in the American Midwest, for instance, where I was born and raised, have become quite wealthy growing corn for biofuel instead of producing grain. Consequently, the price of food has risen for some of the world's poor. Sugar planters in Brazil make more money from the biological waste in their fields than from the sugar contained in the cane plants. They are encouraged to use more toxic chemicals on the sugar lands and to clear more tropical forest to grow it—all to produce more biomass energy. Meanwhile, visions of energy-laden switch grass swirling in the breeze fuels the imagination of many biofuel advocates. But just as the numbers don't add up for everyone to burn firewood in their woodstoves on Red Lick, so, too, the promise of biofuels is hard to envision when we scale it up for the planet. Biomass energy sounds good and green. If we actually were to use plants in such numbers as to fuel the modern global economy, though, Earth's flora couldn't keep pace and our planet would no longer be so verdant. In the short term, we might play musical chairs by shifting land use from food to fuel and back again, but in the long term, we'd have to replace our forests and farms with fields of grass, corn, bamboo, or some other kind of biofuel—hemp in the case of Kentucky—and this would permanently alter the nature of our world. While some plant species grow rapidly for energy production—bamboo, for instance—they don't provide the kind of diverse habitat that is necessary for a healthy planet.

I'd rue the day I lived inside a biomass energy thicket instead of within the calm retreat of a hardwood forest. I'm tempted by bamboo; I grow several species near my goldfish pond, but I don't see it as a viable energy source. Sometimes, I'll burn culms of waste bamboo in a bonfire in our kitchen garden during the fallow season, but I wouldn't think of burning it in the woodstove inside my house. For one, the water vapor and air trapped inside the bamboo's segmented structure expands and explodes when heated, so that when I ignite a pile of it at night in my garden, the staccato bursts of exploding bamboo sound like gunfire ricocheting among the surrounding hills. I can only wonder what my neighbors must think at these times. Technological advances may one day make biomass a supplemental energy source, but I would be hard-pressed to imagine how it might work on a global scale. A friend of mine recently converted his car to biodiesel using the drippings from his tavern grill, and while this has a certain individual appeal, it doesn't seem to be a reasonable solution for a global energy deficit.

I mainly burn oak and hickory in my stove. These species produce a high concentration of combustible heat when compared to other trees. Still, white oak, which is a high-density energy tree, produces only about fifteen megajoules of energy per kilogram of wood weight. (A megajoule is a measure of energy density; 3.6 megajoules is roughly equivalent to one kilowatt hour of energy.) This compares poorly to the heating value of coal, a traditional stove fuel still used by some people living on Red Lick, which has twice the energy equivalent of firewood. Natural gas and crude oil, meanwhile, produce forty-five megajoules per kilogram. And what about nuclear power? Gadzooks! Uranium has an energy density of a whopping 79,500,000 megajoules per kilogram! Such figures help to explain why fossil fuels, despite their toxic output and non-renewable nature, remain preferred energy sources for the industrial economy and why so many people praise the potential of nuclear energy, despite its obvious safety concerns. Other factors such as profits, convenience, and geopolitics contribute to society's continued dependence upon fossil fuels, but energy efficiency has to be taken into account.

I love the smell of wood smoke wafting through Wolf Gap Holler, but I also know it's one of the culprits behind global climate change. Behind the pleasant odor is carbon dioxide and other greenhouse gases. Woodstove efficiency is getting better, though, and this helps a lot. Burning wood in an efficient stove releases only about four grams of carbon dioxide per kilowatt-hour of equivalent electricity produced. Coal, on the other hand, emits more than two *pounds* of carbon dioxide per kilowatt-hour of electricity, while natural gas and oil produce 1.2 and 1.6 pounds, respectively. That coal has such a high energy density value is immensely profitable for the energy industry and provides most of us with a cheap source of electricity, and this explains why it remains a preferred source of energy despite its high rate of carbon dioxide emissions. Many "clean coal" advocates suggest putting scrubbers on coal-fired smoke stacks in order to reduce carbon emissions. But coal is dirty in more ways than air pollution.

Wolf Gap Holler lies a short distance from the coal fields of Appalachia, where land and human society have suffered historical injustices in service to the energy needs of American society. The coal now in Kentucky is obtained by shaving the tops off the Cumberland Mountains and dumping the toxic overburden into verdant valleys located below the mines. The coal industry refers to these kinds of extractions as surface mining, but it's hard to imagine any longer how the term accurately conveys what is going on in the mountains. Some people call it mountaintop removal, which I believe is more to the point. The mining operations employ enormous earth-moving equipment: giant bulldozers and off-road dump trucks that are bigger than my house. A tire from one of those vehicles would completely encircle my goldfish pond, and a single articulated hauler could remove all the dirt in my garden in a single payload.

The mining machinery peels away layers of bedrock until a mountain finally gives up its coal. Where the giant excavators are insufficient for the job, the miners turn to explosives, mainly ammonium nitrate and fuel oil, the same materials used by terrorists to make bombs. Two billion pounds of explosives have been deployed so far in the hills of Kentucky. The mountain top detonations reverberate among the coal towns, fracturing homes and human lives. The coal industry then dumps the by-products into streams that flow beneath the decapitated summits. It does this multiple times, excoriating the surface of the planet like it was the skin of a tortured enemy, cutting ever-deeper slices into it, until finally a mountain is eviscerated, gives up, and disappears from sight. The emasculated Appalachian Mountains are left then with another barren plateau, and the people who live in the region face a new generation of desolate hills, toxic dust, poverty, and tainted drinking water.

Mountain top removal in Appalachia destroys hundreds of thousands of acres of forest and pollutes more than two thousand miles of waterways. Politicians say this is necessary for the sake of economic development. Mining company officials proclaim in local town hall meetings that the flattened ridges will make good parking lots and sites for strip malls, spawning additional local prosperity. Then the coal owners pack up their machinery, lay off local workers, and move to the next ridge in the Appalachian range. *Go tell it on the mountain:* The coal-producing counties of eastern Kentucky are missing three hundred mountain summits, and their per capita income is less than half the state's average (and Kentucky is one of the poorest states in the nation). Elected officials take note: somewhere the calculations have gone badly askew.

Anyone would see that such extractions have a devastating impact on the landscape and human society if the coal companies weren't so good about planting trees along roadways in buffer lines called "beauty strips." These are designed to block the sight of the blown-up mountains from passing vehicles, so people *can't* see what is happening to our world. We all need to get out of our cars every once in a while in order to properly see the planet, but it might take an airplane ride to view the scale of devastation wrought on the Appalachian Mountains by the way coal is obtained. And it's not just that most of us can't see where the coal is mined; we don't see the coal *at all*. We use the electricity produced after coal has made its way from the mountains through power plants and transmission lines to our homes. It's easy to ignore a problem that has been made so difficult to see.

Some people argue that burning wood for heat is a benign and carbon-neutral technology, that it doesn't release any more carbon dioxide into the atmosphere than if the trees were simply allowed to stay on the ground and rot and decay. While that may be true over the long run, it also is a matter of timing. Burning wood in a woodstove releases its carbon *all at once* through a chimney, rather than by slow decomposition in a forest. A dead tree left to rot in the woods doesn't make good firewood because it has already given up much of its energy to the decomposers. A living one, meanwhile, absorbs carbon dioxide and releases oxygen.

The carbon dioxide in fossil fuels, meanwhile, stays in the ground until it is burned, and then its carbon content, which has built up over many millions of years, is released *all at once* into the atmosphere. This is a much bigger problem than woodstoves: a huge amount of carbon dioxide escapes immediately from its long-term underground sequestration when fossil fuels are burned. The plumes of wood smoke wafting above my home in Wolf Gap Holler dissolve against such a backdrop of industrial-scale emissions.

I've been heating with firewood for a long time and am onto my third woodstove. The current version is the nicest I've owned; it has a glass door that allows me to see the fire as well as feel its warmth. It's a pleasant change from the solid wall stoves I've used in the past, which always heated the house well but were real clunkers to look at. There's nothing wrong with a good-looking woodstove as long as it burns wood efficiently and in a clean fashion. In my travels around the world, I spend a lot of time in places that keep warm by hearth fires—herder's tents on the Tibetan Plateau, stone villages in Nepal, caravansaries tucked into the Indus Valley—and in my experience, a woodstove with a glass front is a much better option for viewing the flames than an indoor fire pit. So many times I felt as though I'd asphyxiate while sleeping in one of the Himalayan homes. The smoke put off by their open fires is meant to escape the room by convection through a hole in the roof, but how well it works depends upon atmospheric conditions, the size and placement of the roof vent, and the type of wood being burned. All too often, the smoke gets trapped inside a dwelling where it builds up to toxic levels causing lung-wracking fits of cough and burning, rheumy eyes. In the long run, it may lead to serious respiratory problems or glaucoma, two of the main health issues facing rural people around the world who eat and cook over hearth fires. Smoke is toxic and we all need to know something about it when we heat with wood.

Living with a woodstove also requires knowing about a forest. It's not like heating a house with electricity, where a person just pays the bills and grumbles about the rising cost of utilities, or with an oil or gas furnace, where a delivery truck comes around once in a while to fill a tank. With those kinds of systems, people are lulled into believing they don't need to know much about where their energy comes from. That's wrong, of course, but it's easy to understand why people might think that way. After all, *we* aren't the ones mining the coal or extracting the oil ... or are we? I recall not so long ago when political rallies in support of fossil fuels ended with the eerie chant "Drill, Baby, Drill!" Looking upon the visages of those chanting that message, I felt as if I was peering into the face of madness. I'm not suggesting the crowd was maniacal — maybe it was, I don't know — but the people certainly looked *angry*.

It seems to me that when we heat our homes with oil or electricity, and put gasoline into our vehicles, which we likely will do well into the foreseeable future, we simply need to be responsible about it: learn where the energy comes from and how it's obtained, and know its impacts on planetary ecosystems and the lives of other people. We cannot simply be concerned with how much it costs when we fill our tanks at a gas station or pay a utility bill. This involves settling fundamental accounts about our life on Earth, negotiating a balance between what we need or want or can afford, and what the planet can provide; in other words, taking meaningful stock of such things. When I heat with a woodstove I have no choice but to think about this. While it may not be sustainable in the long run as an option for most people on Earth, heating with wood at least requires some thought on the part of a person. When I sit by my goldfish pond on a summer morning and look into the forest, knowing that it supplies me with some of the firewood I will use to heat my home in winter, I understand there is a lot for me to learn about. I'm not sure a person heating their home with electricity would gain a similar revelation from staring long and hard at a cold, baseboard heater.

I once took a group of students down the Rio Napo River in Ecuador on a canoeing expedition. We found ourselves on the western edge of the Amazon rainforest. Now *that* is a forest. Its territory is two-thirds the size of the contiguous United States. When George and I look beyond my goldfish pond to the surrounding hills, we see thousands of acres of lovely Appalachian forest. It's always an impressive sight and another reason we feel content living where we do in Wolf Gap Holler. The Amazon Basin, on the other hand, contains *1.4 billion* acres of rainforest. It's the largest single assemblage of plant and animal species in the world: one in ten forms of life on the planet resides in Amazonia, one in five types of birds, and one in five kinds of fish. A single pond in the Amazon rainforest may contain more fish species than are found in all the rivers of Europe, and the total variety of Amazonian fish outnumbers all the fish species in the Atlantic Ocean. A twenty-five-acre plot of Ecuadorian rainforest contains more tree species than all of North America's forests. Such numbers are simply staggering.

The forest around my gardens in Wolf Gap is reckoned to be one of the most biologically rich landscapes in North America. It hosts on average thirty tree species in a square kilometer. The Ecuadorian rainforest contains *eleven hundred* species of trees per square kilometer. The Amazon's biodiversity is so great, in fact, that it's not even known for certain how much life it actually contains. Most of it, literally, is up in the air, in the canopy where it remains out of sight. The Amazon rainforest frequently is called the "lungs of the planet" because it absorbs a huge amount of carbon dioxide while releasing almost one-fifth of Earth's oxygen. It literally invigorates our entire world. Moreover, about half the rain falling on Amazonia is produced by its own trees cycling moisture back into the air through the process of transpiration. Given such high rates of oxygenation and precipitation, the Amazon rainforest comes pretty close to being a self-contained ecological system, a sort of global-scale terrarium, were we to just leave it alone. Clearly, the world *needs* the Amazon rainforest.

Yet we are destroying it at an estimated rate of 2.7 million acres a year (one-and-a-half acres per second). Twenty percent of its native cover is gone and will never return to a primeval state. Such a loss is a concern for everyone because the Amazonian ecosystems operate at planetary scales, but the impacts fall most immediately upon the shoulders of native people who reside deep within the rainforest. They are a wilderness people; if the Amazonian forest were to fully disappear, so too would their way of life. It's already happening tree by tree, tradition by tradition.

An estimated six million persons inhabited the Amazon rainforest prior to the arrival of Europeans in Latin America, but the indigenous population dropped to below a million by 1900. The current number is less than 200,000 natives. There are many reasons for their demise—disease, conflict, and migration—but most recently and devastatingly, it is due to the loss of the forest itself. Without sufficient wilderness, it's impossible to be a hunter or a forager. Notwithstanding their dwindling ranks, the Amazonian natives remain vital to the survival of the remaining rainforest in the modern era. Why? They are its watch-keepers, its truest advocates, "eco-warriors" of the highest order, struggling not for an idea but for their cultural survival. They are the guardian angels of the Amazon because they have a vested interest in its preservation: whither goes the forest, so too its native people. Indigenous people throughout the world are engaged in similar struggles as they see their worlds crumbling before their eyes. The Amazonian forest dwellers neither are engaged in an ideological debate nor embarked upon an altruistic conservation campaign, but they understand that the quality of their lives depends upon maintaining a state of balance in their sylvan world. That same balance applies to all of us, living wherever we do, but it may not be so obvious to those living in a city where the trees mainly sprout from sidewalk containers.

The need to conserve the planet's resources is baffling to some people who live remotely from ecological reality. And the preservation of nature is a wilting rose alongside the populist concerns about economic growth and national security. It's not that modern people care nothing about nature—we get rich, after all, by plundering it to the point of destruction—but one person's view on it may be a world away from another's. The technological arrangements of our world have transformed our thoughts about life and nature. The Amazonian Indians seek to protect the natural environment in order to preserve their way of life. The society in which I live *destroys* it in order to preserve ours. Rarely, though, are we asked to think ecologically, or even humanely, about matters that connect us to Earth. Energy and resources are at the heart of it. We are the planet's super-consumers; there is no equivalent in the history of humankind. And there are more of us around each corner with every passing year. We learn that we must plunder the planet at an ever-faster rate or the global economy and our future prosperity will suffer collapse. We are taught to find a sense of self-identity in our level of consumption and to seek our satisfaction and happiness from it. As our numbers multiply and the global economy expands, the repercussions of the way we live are felt deep within the planet's life-supporting systems—the atmosphere, landscapes, and oceans—rippling through the geo-physical cycles, making inroads across plant and animal kingdoms, and ultimately coming home again to roost among our societies, and, ever-closer, in our hearts and spirits, where we must look for hope. Is our way of life truly so dependent upon the destruction of our home? Is that what so many people believe? Would it make sense for me to rip up my gardens, empty my goldfish pond, and cut every tree in the forest just so I might live a more prosperous life? How do we take stock of such things?

I took students to the Amazon on an educational adventure. We stayed in a rainforest village of Quechua-speaking Indians, one of the Amazonian tribes. There we spent much time among their company in the forest. It was obvious that they were at home in the jungle. It was equally apparent that we were casual visitors. They ran lightly among the trees, flitting along the trails like Blue Morpho butterflies (*Morpho peleides*), while we tripped and stumbled over roots and vines, and plunged into slick, muddy holes. They wore simple loincloth; we wore knee-high boots and tight-fitting clothes to ward off the biting insects, and, as a result, sweated profusely much of the time. They saw birds, butterflies, and reptiles, but we saw nothing but dense entanglements of undergrowth. It was a frustrating experience for many of us.

Our days began with conversations in the early cool hours when the students sat in a small clearing on folding chairs. These mini-seminars prepared us for what we might experience that day. One morning a villager found a poisonous fer-de-lance snake coiled directly beneath one of the set-up chairs. A few moments later, a student would have taken a seat there, unknowingly dangling her bare ankles directly in front of a pit viper. Another day, a villager extracted a python from the rafters in a student dormitory room. One time we went into the forest looking for birds, accompanied by an ornithologist and carrying a book of glorious photographs of the avifauna we hoped to encounter, but we never saw a single bird. The jungle was alive with them, of course, but the vegetation was too dense for us to spot them in it. We could hear them flitting all around us. Even our native guides had difficulty seeing them, but they could identify the birds by their whistles, screeches, and songs. We couldn't do that, either.

A week into our expedition we got stranded by civil disturbances taking place elsewhere in the country. Stuck in a tiny lodge in a remote village, we soon ran out of tourist food and were forced to eat starchy cassava several times a day: cassava porridge for breakfast, fried cassava or cassava soup at lunch, and cassava pizza at night (this was a student group). A few days into this tuber diet, the students were climbing the walls. They grew bored and restless. I was concerned they might go bonkers: strip naked like the actors in an old Tarzan movie, paint their bodies with ritual markings, wear cockatoo feathers in their hair, plunge into the forest never to be seen again. What would I say to their parents? Students need food. Fortunately, the local villagers eventually augmented our cassava diet with tasty and nutritional foods they gathered in the nearby forest: fried fish caught in the river, forest greens in a salad, and plates of delicious tropical fruit. All we had to do was ask them for help.

The forest knowledge of the Indians living in Amazonia is far beyond what someone living in Wolf Gap Holler might reasonably need to know about the Kentucky woodlands in order to heat with a woodstove. Still, some people in Wolf Gap know more than others about the forest. Some of my neighbors know most of the tree species. Others can identify wildflowers in a forest just as handily as if they were cut blooms in a vase on a kitchen table. My brother-in-law has a keen ability to spot edible mushrooms. Morels are the most famous of the local fungi. They're found at specific places within the forest and sprout from the soil in the springtime when the soil temperature and moisture are just right. A person needs to be in the proper spot in the forest and on the right day in order to find a morel mushroom, and even then it's difficult to see one: a morel gatherer must look carefully among the decaying leaves and other forest debris. A lot of people go into the Wolf Gap forests looking for morels, but only a few will return with enough for a decent meal. In some parts of Kentucky, people call them *merkels*, presumably named after a frontier family, the Merkels, who allegedly survived a desperate season in the mountains by eating morels, but I think the mushroom got the name *merkel* because it is so challenging to find one that it's close to being a *miracle* when it happens.

Some folks gather wild ginseng in the Wolf Gap forests to sell to local herb traders who, in turn, market it to Asian importers. The Chinese consumers value Appalachian ginseng over their own domestically grown variety. Ironically, this connects the forests of Wolf Gap Holler to the boom-time cities located on China's eastern coastline. Ginseng is fast-disappearing from the Appalachian woods because it's being over-harvested for export to Asia. A lot of local people also hunt game in the forest and know quite a lot about the behavior of wild animals, mainly deer and turkey, but also squirrel, rabbit, and the occasional raccoon. Sometimes we consider hunting the family of whitetail deer that graze at night in our gardens, but that would require shooting them from a stand in the magnolia tree near my goldfish pond or from the rooftop of our home, neither of which seems quite right, and it becomes an even more distant prospect whenever we see a spotted fawn. Bambi-like in the garden, but imagine it being alone and without its mother while waiting anxiously at the edge of the woods.

All in all, people in Wolf Gap Holler who heat with a woodstove share a pretty basic knowledge about the forest, regardless of whether we buy, scavenge, or cut firewood ourselves. Time and inclination determine how far any one of us will advance in our knowledge about the natural world. We live in different ways, with diverse sets of abilities and inclinations, and not everyone can devote a lifetime to the study of a forest. At minimum, heating my home with a woodstove gives me a chance to turn on its head the old saw, "it's hard to see a forest through its trees." The popular phrase suggests that it's difficult to discern an overall pattern in the world because we are distracted by its details. In fact, it is *because* I take an occasional tree from the forest around my goldfish pond and burn it in my woodstove that I become more aware of how I fit into the overall scheme of life on Earth. ❖

CHAPTER ELEVEN

I've read a few books about garden ponds, but most of what I've learned comes from trial and error. This sometimes is unfortunate for a fish or a plant, and even a frog, because I make mistakes and the outcome might be fatal for them. For instance, the water chemistry may be out of whack, or I might have upset the energy balance by skipping a beat in the trophic level of the pond. I hope to eventually learn from my mistakes, but in the short term they can be matters of life and death for a frog or goldfish

What Is There to Know?

Several years ago George and I traveled from my goldfish pond in Wolf Gap Holler to southern Georgia, where he was born and raised. I was keen on visiting an alien landscape while George was heading home to his roots: black water river and cypress knees, a tiny cabin by an alligator pond. Motoring on a back road not far from George's home town, we passed a group of men lounging beneath a shade tree. Some of the guys were seated comfortably on fold-up aluminum chairs. Others stood in relaxed poses or sprawled lazily in the dry grass. They appeared to be engaged in some kind of friendly confabulation. A scrap of plywood tacked onto the tree above the men proclaimed "Tree of Knowledge." The hand-lettered signboard piqued our attention, but we didn't stop to converse with the men. We slowed down and considered it, but then kept driving down the road. In hindsight, we probably should have halted, gotten out, and asked those men: *what is there to know*? We missed a golden opportunity.

George and I sometimes dispute knowledge and learning when we sit by my goldfish pond in the morning. It's a kind of shop talk for us. We both are professional educators, having taught for several decades at colleges and universities, although George now is retired and I'm soon to follow. From where we sit, we see the developments in the education field at a bit of a distance. We watch resources being poured into schools with little apparent result, as if all the money and new programs just keep the educational system from slipping backward at a faster rate. And it's difficult for many teachers to know for certain what students should learn these days, the world is changing so fast. This is especially troublesome on a college campus, where the traditional curricula are routinely discarded for new ones targeted at the rapidly evolving job market. It's understandable—education needs to stay relevant—but the changes happen so fast now that a single class of students might be caught between competing curricula, and then it's even less clear what they need to know before they graduate.

In the absence of state funding, tuition is what keeps colleges solvent these days, so much of the curriculum in higher education is aimed at recruiting and retaining students rather than the broader needs of an educated civil society or the old-fashioned rewards that might come to a person with a liberal arts education. The public education well, based on increasingly reluctant tax payers, is running dry, and so students must dig deeper into their own pockets to pay for their college education. Tuition will continue to increase under the new financial arrangements, which means that a college degree will be further equated with a person's lifelong earnings potential. It's a slippery slope where knowledge is tied so closely to a student's prospective income and is seen as less of an investment in society or a person's capacity to live a thoughtful life. The value of knowledge is tricky to measure and is further obscured by these calculations of profit.

My students already know the power they wield as consumers, having been the targets of marketing campaigns their entire lives, and many have no reservation about taking their custom to the university registrar. Many of the student-centered initiatives I observe on campus are designed to appease them, so they don't go and shop at another university. This assumes, though, that students know what they need to know when many only know what they want, as if they were in a shopping mall. Information has become the world's most valuable commodity, but it's not the same thing as knowledge. We can easily find information; we live, after all, in the *Information Age*. It's for sale everywhere and drives the global economy. My college students are on the frontlines of the information frontier. They know more about it than I ever will. Knowledge, however, involves more than retrieving data or news. It may seem unreasonable these days when universities are becoming information malls, but how on Earth can we expect a young person to learn what they need to know in life when we just give them what they want? Sometimes I feel like yanking those guys out from under their shade tree in South Georgia, haul them up to Kentucky, and give them the job of running the state's education system.

I've read a few books about garden ponds, but most of what I've learned comes from trial and error. This sometimes is unfortunate for a fish or a plant, and even a frog, because I make mistakes and the outcome might be fatal for them. For instance, the water chemistry may be out of whack, or I might have upset the energy balance by skipping a beat in the trophic level of the pond. I hope to eventually learn from my mistakes, but in the short term they can be matters of life and death for a frog or goldfish. I've always found that field-based learning, whether it's inside a garden or out roaming the world, is invaluable because it speaks to the essential ingredients of knowledge: imagination, experience, and discovery. And it's one of the reasons I travel so much: the planet is a world-class school of learning opportunity. Moreover, I teach geography, and what better subject is there to study when one is off exploring distant places? I try my best inside a classroom, it's my job, but the view out a window always falls short of the distant horizon. It's not like we all have our minds inside a clothes dryer set on the spin cycle. I can't time an *idea* like I can a load of laundry. Yet, that's what it feels like much of the time inside a classroom, a spin cycle running at top speed with the contents going round and round in our heads for the duration of a class period, an academic semester, or even a life. There's little room in this approach for contemplation, without which critical and creative thinking go down the drain with the suds and gray water. It may be that in the years to come, for my students who will then be adults, the only speed that will make sense to them is cyber-speed. Moving that fast, though, one might not see the learning opportunities that await us in the quiet corners of the world, or even notice a pretty little goldfish pond in a country garden.

I've been with students in some pretty amazing places — slogging through wet jungles in the Amazon, climbing among snow-topped ridges in the Himalaya, swimming on coral reefs and hiking up volcanoes on remote islands in the South Pacific — but I've never thought to bring one into my Wolf Gap gardens. I know there's a lot to learn in them, but I see my gardens as private enclosures and not public school classrooms, a place in which I'm the student and not a teacher, and, besides, any knowledge they might offer would require a person to have a certain familiarity with nature. A college degree in horticulture isn't necessary, but the precognition that comes from willingly opening one's heart to the natural world is surely a class prerequisite. I suppose my gardens are best suited for students who have already done a certain amount of homework.

Because my gardens result from a life of travel, in addition to the plodding sort of tillage I practice at home, my gardens have meaning for me beyond their immediate appearance. A garden may not readily give up its intentions and one can't expect it to. I find it's when I'm paying attention to the details of a garden, its symmetry, textures, and timings, and being cognizant of its overall balance, that I can most benefit from a walk in a garden. A person has to dig deep for garden knowledge. Since it's an exceedingly pleasant way to pass the time, we can *enjoy* a stroll down a garden path without any of that, but we might not learn a great deal from it. We have to move slowly and center into a garden for knowledge to arise from within it. A garden is like the rest of a person's life in that way. If we want to learn something useful about the world, we need to engage with it and live in ways that are both deep and wide. It is for these reasons and not solely because of my wish for privacy that I haven't yet turned my gardens at home into a classroom for my students.

When George and I sit by my goldfish pond, we find pleasure as well as verity in doing so. Despite all outward appearances, we're not just sitting idly by. One of the things I find most tiresome about formal classroom instruction is that learning is construed in ways that are no fun at all. My students appear glum much of the time, as if they wished to be somewhere other than in a class room. Jennifer, meanwhile, couldn't *live* without being in her gardens; they are that instructive and enjoyable for her. Our cat Olive thinks that we are making the gardens just for her to play in and to learn about the world. Olive got lucky when Jennifer picked her up one cold February morning from the side of the Red Lick Road. She was a little lost kitten. Now she's learning about life while frolicking among our gardens. In fact, Olive knows no other world. Apart from a few tentative explorations into the surrounding forest, her world *is* our gardens.

I often play with the cat in my Oriental Garden, to which she has taken a fancy for some non-metaphysical reasons. She darts from rock to rock, bats at leaves and moss as if they were moles, scoots beneath shrubbery, and glides along the rickety tops of the bamboo fences like a ninja fighter. She leaps onto a squat pagoda beneath a black pine, using it as a perch to survey her surroundings, and then launches herself into the mondo grass and flattens her body against the ground, hiding as though she were on a hunt; perhaps in her mind she is stalking blackbuck on the African savanna. At such times, Olive is in full "play" mode: she's alert, frisky, and in full engagement with the world. I enjoy playing with Olive when she's in this state of mind because ... how shall I put it ... she's *amusing*.

Of course, I have no real idea what's inside the mind of my cat, just as I often have no clue what my students are thinking. This becomes apparent whenever I find one of them surfing on the Internet instead of paying attention in the classroom to the matters-at-hand. I think it's laudable that my students want to surf, but then I think they should go elsewhere, like Hawaii, and learn something useful by riding a wave in the ocean. Not all places are tuned in the same way, and we should try to match what we are doing with where we happen to be in the world. When my cat and I play in the garden, we seem to be thinking along much the same lines—how to outwit the other, a sort of "hide-and-seek" game—and so, to a certain extent, what's in her feline head is the same thing that's in my human mind. A garden setting is conducive to this kind of shared introspection. I wish I could say I frequently have such experiences with my students in a classroom, but it's not all that common. I sometimes wonder whether or not I have a more meaningful intellectual relationship with my cat than with some of my students. What does this say about my students, my cat, or me?

If it's hard to decipher what is happening with students on campuses these days, it may be due to all the confusing slogans flapping around schools and colleges. We have the No Child Left Behind policy and the Preparing Students for Success initiative. We have the Students Come First program, and the Education is our Business slogan, which I believe gets right to the point of higher education trends. We have banners on campus buildings, coffee mugs, and tee shirts, all proclaiming Building a Better World One Student at a Time, or Education: Your Door to the Future, or Soaring to Success. We have school mastheads inscribed with: A Great Place for Education; and, on my campus, an entire facility is named EDUCATION PAY$. It's as if the marketing department has taken charge of the education system, yanked it out from under the guidance of its teachers, and put in place a set of snappy new slogans, thinking, perhaps, that a new brand will solve the learning problems of America's youth and lights will pop on inside my students' heads. It's a mind-boggling proposition.

One year I taught a group of American college students in Nepal for four months. It was an amazing semester for me, one of the best in my career as an educator, and it was mainly due to all the things that went wrong and from which we learned a great deal, not so much about geography, although that, too, was there, but about life in general. I am indebted to Nepal for providing me with so many opportunities for adventure and discovery over the course of my life and was grateful to be able to share some of that with my students. We began the semester with a three-week trek to remote villages located high in the Himalaya. It was a rough time for some of the students, living close to the bone, and I was pleased that they were able to hang in there given the circumstances: strenuous hiking, monsoon rains, no electricity or running water, no cell phones or Internet, no pizza, a lot of frigid nights. About a week into our journey, we heard that a plane had collided with a building somewhere in the United States. It was the evening of September 12.

A short-wave radio located in one of the villages broadcast bits and pieces of news, but it was hard to discern what had happened. Some of us thought the Sears Tower in Chicago was hit by an errant aircraft because it's so tall and seemed a likely target for a disoriented plane lost in the Lake Michigan fog. The following day we gained better communications and learned more about the tragic event of September 11 — not the full story, but enough to know that something catastrophic had occurred in America. The news of it sent my group of students into a tailspin. Some of them lived near New York City, within sight of the World Trade Center, and others had family in the city. A few students knew people who worked in the Twin Towers. We had no telephone in the mountains, so I sent a runner to Kathmandu to gather more news and to make contact with the students' families. Much was involved and some days passed, and in the end, one person opted to return home, while the remainder, having learned their families were safe, continued the journey into the mountains. Onward student soldiers! It was a difficult call to make — the emotional toll, complex logistics, and sheer physicality of the place were trying — but the students decided to move forward, understanding, I think, that the horrible events of September 11 were already in the past and half a world away.

A few days later, I suggested to the students that we climb a nearby ridge for views of the big summits. I thought it would be therapeutic to see the snow peaks, to clear our minds somewhat of the unsettling events that had occurred in America, and to center ourselves back into the Himalaya, or we might never learn anything by being there. The monsoon season was drawing to a close and cloudy skies had accompanied us during our trek up to the high-altitude villages. We hadn't seen a single peak the entire time, although the glistening peaks were all around us, hovering over my emotionally scarred group of students like unseen sirens beckoning us ever-closer. We had walked for days in the rain on soggy trails and slept in cold villages under leaden skies. The students' knowledge of the wondrous topography ahead was limited to my claims or to pictures they'd seen at home. It was time to pay a visit to that mythic world of plunging valleys and snow-clad summits.

The weather was changing on the turn of the seasons. The days were dawning clear but stayed fair for only a couple of hours. By mid-morning, a flow of moist air swept up the valley to form clouds that swelled like billowing armadas set sail for the mountains. Our little expedition had put us at the base of the summits just as summer was giving way to autumn, and the meteorological changes signaled the desinence of the rainy season. Since the window of clear skies lasted only a couple of hours, we would have to climb early and fast if we hoped to beat the clouds to the top of a 16,000-foot ridge. From there, we would be able to see the magnificent landscape that straddles the border of Nepal and Tibet. We left our village in the dark for the rough ascent.

At times, I felt like a drill sergeant hustling a bunch of new recruits up the side of Bunker Hill. The students were anxious to see the mountains, having traveled so far from their homes, and their excitement carried them for a ways up the trail. Soon, though, one by one, they flagged as their energy sapped and exhaustion set in. I prodded, cajoled, and even lied to some of them, suggesting we were almost there when we were not yet close, in order to keep everyone moving along at a steady clip. Midway up the mountain the group splintered, with some students halting to rest while the others kept going. The weary would catch up with the rest of us after a short break. Another hour into the ascent and the higher climbing party looked down the valley to see clouds taking shape at the low elevations. The rising nebula would advance quickly with the lifting air and gave urgency to our climb. It became obvious to everyone that we would have to move even faster if we hoped to gain a view of the mountains before they were obscured by the clouds.

We were climbing amid the powers of the planet at play: geological uplift harnessed to precise atmospheric clockworks. And we had to synchronize our efforts to this great turning of natural forces. The diurnal arrival of mist-laden winds was as regular in the mountains as tides in an ocean. We couldn't hope to stop or slow down the movement of the clouds any more than we could halt the drift of continents that give rise to the Himalaya or relieve the pull of gravity that made it so difficult for many of us to climb the ridge. We moved forward in the face of tectonic uplift and the Asian monsoon. We were racing the clouds. You can't get *that* kind of geography lesson in a classroom on a university campus.

We finally gained the top of the ridge and looked out upon a crenelated landscape of vaporous gorges and up-thrust rock walls. The earth appeared to have convulsed, and its seismic shudders suddenly frozen into rock and ice. The snow summits loomed above us. We saw that other people had been there before; the ridge was a holy place and had been visited by religious pilgrims for many centuries. Fragments of prayer flags stretched in a line along the exposed ramparts and flapped in the elemental colors of Buddhism. A cobalt blue sky embraced the icy mountains. We sat on a ledge, transfixed by the view, and watched as the clouds caught up with us, at first in fleeting, opaque drifts and then gradually forming a thick, impenetrable wall of spume that obscured the mountains, leaving us stranded in a milky void. A few minutes later, the students who had halted below gained the top of the ridge, but by then the vista was gone, vanishing without a trace into the monsoon-laced opalescence.

We waited in the swirling mist. I asked the students if it had been worth it. Had it been too much trouble to climb in the cold and damp when at best we could hope to see the summits for only a few minutes and some might not glimpse them at all? I thought they must be tired, disappointed, and frustrated, thinking I had somehow tricked them into joining me on some kind of farcical expedition. Their answers, though, were heartening. Somewhere along the trail, the view had become secondary to the effort of getting to it. They felt good about that. Our lessons about mountain geography made a new kind of sense for them. They felt in their fatigue a measure of Earth's power. Our climb against the pull of gravity toward a distant ridge had pushed the horizons of their own appraisals of the world. We hadn't tackled a big mountain, of course, not the kind of peak that an alpinist would brag about climbing, but it was *our* little summit: a ridge of adventure, wonder, and hard-won inquiry. The Himalaya is filled with such places, which is why I keep going back, and seated that day on a ledge with my students, high above the void, observing mountains and rising clouds, noting their reactions to the events unfolding around us, I learned again how much wisdom Earth has to offer when we ask it to be our teacher. ❖

CHAPTER TWELVE

When the frogs croak amid the pond lilies and fireflies glow in the evening sky, it seems to me that the world is about as beautiful as it can possibly be.

Twilight at the Goldfish Pond

A few events transpire at the end of a summer day to herald evening at my goldfish pond: the sky darkens, stars begin to shine in the welkin, pond frogs sing, and fireflies sparkle amid the shrubbery. They signal the commencement of a *son et lumière* show in my garden to rival the grand productions staged for globe-trotting tourists at world-famous monuments. I've seen them in India's Red Fort, among temple grounds in Thailand, at the Greek Parthenon, always enchanted by their lavish displays of light and music, but I've come to prefer the little shows in my Wolf Gap garden. Free and improvisational, installed without fanfare, the performances at my goldfish pond require neither road crew nor truckloads of heavy equipment nor fancy proscenium stage: it's theater-in-the-round in my garden, a setting that changes with the seasons as plants burst forth with new foliage and flowers bloom and wilt amid the biogeochemical cycles that spin on a fulcrum of seasons. These frog-and-firefly entertainments take place in my garden on most summer nights — the show must go on — and I'm glad to watch whenever I have the opportunity.

Knowing a few things about the cast at the goldfish pond theater adds interest to the show. The lifespan of a firefly, one of the luminaries in the cast, averages a hundred days, which means they light the performances at my goldfish pond for most of a season but never longer than that. Frogs, on the other hand, whose croaks add a musical element to the shows, may live ten years or more, and so it's likely they're old hands at these garden shows, showing up year after year to play in the band like dues-paying musicians at the Grand Ole Opry. The duration of stars in the sky above the pond, meanwhile, lies far beyond my comprehension. However, it's safe to say they will be there until the galactic end. Many of us are accustomed to the light and sound systems used in the high-tech *son et lumière* shows, but what do we know about the candela of a firefly or decibel of frog music gracing a pond performance in Wolf Gap Holler, Kentucky?

A firefly lights up because of something the ecologists call bioluminescence; they contain highly-specialized cells and enzymes in their abdomen that create light through a chemical reaction. The light travels in wavelengths of about 600 nanometers, which our eyes detect as a pale yellow to reddish green color. This is why, when I watch the fireflies sparkle in the sky above the pond, they appear in slightly different hues, as if I am seeing pinpricks of luminosity fly like sparks off Earth's molten core. The lightening bugs glow to attract and seduce mates, not to entertain me, and I doubt they'd know I was watching unless I were to catch one in a jar, like I did as a kid, and peer at it through the glass, but then I think its light would grow dim without the prospect of enticing a mate.

Biologists have measured the light emitted by a firefly to be approximately one-fortieth of a lumen, which equates to about one-four-hundredth of a foot-candle of light. As an equivalent, a person requires at least forty foot-candles to read a book without producing eye strain. So I would need about 160,000 fireflies kept in a jar to sufficiently illuminate a page for perusal in a garden. I'm not suggesting that I would actually do that, read a book in the garden by the light of fireflies, for it is a lot of fireflies to keep in captivity. Moreover, fireflies don't light up all at once; they sparkle. Considering this, it's impossible really to calculate the quantity of fireflies needed in order to read a book by their light, certainly more than I could count. It's best, then, for me to leave the fireflies twinkling in the sky where they belong and where they can hope to attract a partner, and use a flashlight if I want to read at night in my garden.

The sound produced by a frog when it croaks is easier for me to comprehend than the wattage of a firefly, but it, too, is complicated by the fact that the levels vary by species. The common tree frog, for example, emits eighty-five decibels in a single croak. That's about as loud as city traffic heard from inside a car when the windows are shut. The forest around my pond is filled with tree frogs, and once they all get going it sounds like rush-hour in a busy Third World city. Some frogs croak at a volume greater than one hundred decibels. That's louder than a hand drill, which clocks in at ninety-eight decibels, and roughly equal to the sound of a snowmobile or motorcycle. It's getting close, in fact, to the one hundred fifteen decibels put off by a rock concert. So when the frogs are singing in my goldfish pond, in full accompaniment to the fireflies sparkling in the sky, it can make quite a cacophony in the garden. But just as fireflies sparkle to seduce a mate, the frogs are croaking mainly to attract members of the opposite gender; they are not serenading me. Furthermore, only the male frogs croak and sing in a chorus. Female frogs make noises, but it's more of a cajoling whisper. And being territorial creatures, the male frogs also croak to warn other frogs to back off, so I have to consider the fact that when I listen to the frogs singing at night in my pond, it may not be a love ballad I'm hearing, but an argument.

While differing in their musical abilities, frogs of both genders are talented jumpers. George and I often witness this during our morning interludes of coffee and conversation in the garden when my cat joins us at the goldfish pond. The frogs, rightly suspicious of her feline intentions, immediately jump into the water at Olive's arrival and sometimes leap from lily pad to pad to get across the pond, so urgent is their desire to escape. That always is a remarkable sight. Jumping frogs have been entertaining people for a long time, not just George and me in Wolf Gap Holler. Mark Twain wrote a famous story about them in 1865 called "The Celebrated Jumping Frog of Calaveras County." His frog protagonist was named Daniel Webster, and its owner had put a wager on how far the frog could jump in a competition. Daniel was fed some buckshot by another frog gambler, which weighed it down, and as a result, the frog lost the jump and its owner the bet. I don't name my frogs, nor place bets on them—frogs aren't horses, after all, and I live in Kentucky, which is horse-racing country—but I gain a lot of pleasure in seeing one make an especially good vault across an open stretch of pond water.

The distance a frog can cover in a single bound is astonishing. The Australian rocket frog, for example, can leap fifty-five body lengths. The world record for a single frog hop is thirty-three-and-a-half feet, which is equivalent to a person jumping half a mile. If I actually were to bet on my frogs, like some Kentuckians do on their horses, and the frogs were equipped with saddles and reins, it would be quite a gambol for a jockey and gamble for me. Frogs spring to escape danger, not to win a wager, and sometimes they leap right over one another in the excitement of a chase. I've seen this only once. George and I were at the pond one morning when a pair of frogs began leapfrogging from one lily pad to another in a great hurry to get across the water. It was a lot of fun seeing them at it. I had played leapfrog as a child but didn't know frogs actually indulged in it. I thought it was just a good name for a kid's game. American children aren't the only ones who play this sport. The French version is called *saute-mouton*, which means "leap sheep." It's known in India as "horse leap." In Romania, it's called "mounting rack" and in Korea "pile driving." I don't know about the Romanian or Korean versions, they sound kind of ominous, but the American game has kids vaulting over the backs of one another in order to get ahead, just like the frogs did that morning in my goldfish pond.

I'm usually the only person at the pond in the evening when the *son et lumière* show commences, so it's always easy for me to get the best seat in the house. At these times, I don't even have to share the bench with George. The show gets going just as dusk settles onto the garden and pinpricks of starlight appear in the evening sky. A few isolated and tentative notes drift up from the aqueous pit of the pond as the frog band slowly warms up. When they get into a groove—get their frog mojo moving—the music really begins. It's not like in the morning when the frogs are just tuning up; they've had the entire day to practice and in the evenings come onto the garden stage like true professional musicians.

Don't get me wrong, it's not that I think the frogs are singing a cappella or performing the complicated movements of a Baroque string quartet in the pond. I doubt the frogs have heard of Bach, Vivaldi, or Handel. I don't think they care much about harmony or counterpoints, and probably can't be bothered with technical matters like functional tonality. They are, however, born masters of dissonance. They flirt with the traditional boundaries of musical genres. In a classical vein, the frogs might feel at home in the Viennese school, performing the syncopated arrangements of composers such as Schoenberg or Varese. But I think the frogs really are the adepts of heavy metal, following the likes of Iron Maiden or Metallica, which seem to me to be more their style, rather than stuck in a stuffy Victorian parlor playing for an audience wearing ruffled collars and powdered wigs. Sometimes they sing and croak in the free-wheeling and edgy way of Miles Davis blowing his trumpet in a blues rhapsody. The frogs are protean performers and like to mix it up at twilight in the goldfish pond.

One might imagine it would be difficult for the fireflies to choreograph their dance in the sky to the croaking cacophony coming from the frogs afloat in the orchestral pit of the pond, but they have no problem at all in doing so. This is because they sparkle. A twinkling light can be timed to just about anything, even frog music. Imagine holiday lights at Christmas, blinking green and red to the sweet chorus of children singing carols. Or, the twinkle in a smiling person's eye: clearly such a one follows a life guided by song lines.

When the frogs croak amid the pond lilies and fireflies glow in the evening sky, it seems to me that the world is about as beautiful as it can possibly be. It is in a state of balance, the Balinese concept of *alango*. At the risk of over-stretching, I would even say that when the fireflies sparkle among the stars, such that the two become interchangeable as tiny holes into heaven, and the frogs sound like tantric lamas chanting monastic incantations to the rhythms of Thelonious Monk, while I sit quietly by, the *son et lumière* show at my goldfish pond is yet another harmonic wave passing through a center of the Universe.

Once a month the frogs and fireflies at the pond are joined by the full moon. This is Act Two in the pond theater, when the luminous orb arrives as mountebank in the garden show. At this time, the moon, big and round, full of itself, rises from the East to burst onto the stage with an incandescent power that dims the stars overhead, so that slowly, one by one, they disappear as if yanked offstage by someone in a top hat and with a hooked cane. It's not that in Act Two the moon simply makes its appearance in the night sky; the moon now is in the goldfish pond, at the center of the show, and its specular reflection oscillates gently in the ripples as if they were a mirror onto its own librations. The lunar orb takes over as star of the garden show by the immutable laws of light and physics. Yet the reflection in the pond limits its appearance to the super-stratum of water. It fails to penetrate the depths, remains incidental, and in full reversal of itself hanging in the sky.

The stars are no longer the luminaries they were before the arrival of the moon. Even the fireflies lose some of their luster to the mirrored face of the moon. Everything now revolves around the satellite. It's a show-off performer, a pampered celebrity floating in the middle of the pond, acting like one of those Hollywood types we read about in the tabloids at the check-out line of a supermarket, the one who shows up late at a pool party with a deep tan and all heads turn in her or his direction. The moon in the goldfish pond is a false pretender and occupies center stage at the Wolf Gap *son et lumière* production only by deception and illusion; as they say in show business, it's all smoke and mirrors. Even so, the fireflies and stars, bright as they are, can't really compete with the lunar glow in the pond; they're too delicate and ephemeral to be clearly seen in the bouncing light of the moonlit water.

When I see the reflection of the moon in my goldfish pond, I'm reminded of a story I once read about a famous Japanese gardener who was instructed by an emperor of Japan to design a garden that would prominently display Mount Fuji. The Japanese are some of the most inspired gardeners in the world, and this old man was the most learned in the archipelago kingdom. The Emperor had his sights set on a grand garden. Japanese horticulturists practice a sort of geographic mimicry by using a technique called "borrowed scenery." This man was instructed to "borrow" Mount Fuji, a huge volcano looming over Honshu Island, and bring it into the Emperor's garden. "Borrowing scenery" requires that a garden include landforms, rocks, and plantings that resemble elements of the surrounding natural landscape. It fools the eyes and tricks the brain into thinking that something is other than what it appears to be, like a house painter employing the technique of tromp-l'oeil on the walls of a home so they take on a three-dimensional look. With proper scale and perspective, a well-designed Japanese garden enclosure will appear to stretch far beyond itself. A gardener might prune conifers to mimic a distant grove of pine trees, place lichen-covered rocks in the garden in such a way as to suggest they tumbled down from a nearby mountain, or install a small water feature in such a way as to appear to be the tributary of a much larger stream flowing just beyond the garden's perimeter. Borrowing scenery makes a garden seem boundless, just as the moonlight and stars above my goldfish pond extends my garden to the Empyrean. But as accomplished as a Japanese gardener may be in doing this sort of thing, installing a four-thousand-meter volcano in a small enclosure is no mean task.

Moreover, this gardener was charged with bringing not just any volcano into the garden, but Mount Fuji, the most revered mountain in all of Japan. It is the country's highest summit, a symmetrical, snow-capped cone of renowned beauty, a sacred mountain visited by thousands of pilgrims each year and enshrined in some of the finest works of Oriental art. This ruler wanted it displayed prominently in his garden, to be its centerpiece. After giving his orders, the Emperor left on a long campaign of war and plunder. When he returned to his palace after many years abroad, the gardener took the Emperor on a tour of the completed garden. It was by all measures an extraordinary achievement, the masterpiece of the aged gardener's long and distinguished career. The garden had an exquisite arrangement of plants and rocks, polished stone pathways for strolling about the grounds, and a secluded tea house for private rendezvous. It was enclosed by a beautiful stone wall and groves of luscious, golden bamboo. Yet the Emperor was incensed for nowhere in the garden could he see Mount Fuji. He called the gardener to task, berating the elderly man and threatening to chop off his head in penalty for this omission.

The gardener waited until his majesty calmed, knowing a person needed tranquility to understand a garden, and then led him on a guided tour of it. They strolled together on a flagstone path that meandered toward a roughhewn bench set above a tiny pond in the middle of the garden, where they took a seat. The Emperor sat silently fuming on the bench. The gardener patiently sat beside him. As the ruler relaxed, he looked again at the garden, took in carefully measured proportions, pruned trees, imaginative rock ensembles, and ways in which the foreground cleverly mirrored the distant hills. He peered into the bamboo grove where colorful birds flitted and sang. He turned his gaze to the small disc of water that lay directly beneath where the two men sat and saw the reflected face of Mount Fuji. The Emperor smiled and his companion sighed. The gardener had brought the volcano into the center of the garden, as had been requested, by immersing it in the small pond, just as if he had set a lovely cherry tree blossom into a bowl of holy water.

I'm always glad to be reminded of that story. When I'm in my garden and the moon is big and bright, outshining everything else in the sky, I miss seeing the stars, although I know they're still dangling above me in the galaxy. I'm a casual sky-gazer. I can't seem to remember one asterism from another, excepting the Big Dipper and Orion's Belt. Those two star clusters are obvious even to me, and a small child will pick them out in the evening sky with just a little coaching. The more obscure formations in the galaxy that relate to the signs of the zodiac lie beyond my comprehension. I don't fret over my ignorance, though, I just like looking at the night sky and wondering about the stars that fill it. I'm not alarmed by the fact that I can't name individual stars or follow their trajectories across the heavens, which anyway change with season and latitude so that I would always be prone to confusion. I can't decipher how the stars align to Earth's magnetic declination. I'm not any good at trigonometry. For these and other reasons, I would not make a good celestial navigator sailing across the world's oceans. Ironically, though, some of the finest star-gazing in my life has been on the islands of Polynesia, which also are home to the greatest star finders in the history of humankind.

One time I was camping in the caldera of a dormant volcano on the island of Maui and devoted my nights to looking up at the stars. From my tent tucked deep within the cindery bowl of Mt. Haleakala, the stars shone with otherworldly light. Haleakala means "House of the Sun" in the Hawaiian language, an appropriate appellation given the volume of sunlight caught inside the crater during the day, but after spending several nights in its sky-piercing concavity, I thought the volcano might also be known as the "House of the Sun and Stars." The starlit sky over Haleakala at night is every bit as luminous as the sunshine that fills the crater during the daytime hours. The Polynesians arrived on Maui more than a millennium ago from the Marquesas Islands after sailing for weeks in their voyaging canoes. The mariners employed star navigation to find their way to Hawaii, measuring latitude using Polaris and the Southern Cross, fixing their ever-changing positions by the setting or rising stars, charting their course within a sacred geometry of known constellations. Additional information gleaned from the sea augmented their celestial observations: changes in water temperature and salinity, the rise and fall of ocean swells, flights of seabirds, promise held in a greenish tinge to distant clouds that indicated land might lie directly beneath.

Looking up at the glittering sky from inside Haleakala Volcano, gazing at stars that had guided the Polynesians to Maui more than twelve hundred years ago, I reckoned I was on a real Hawaiian vacation. The lights of Waikiki couldn't hold a candle to it: no neon karaoke bars, no television sets glowing inside hotel rooms, neither traffic lights nor camera flashes. So dark and clear, in fact, is the night sky that a neighboring volcano on the Island of Hawaii (the Big Island) hosts the densest concentration of astronomical observatories in the world. On the summit of Mauna Kea work some of the most accomplished star-gazers in the world. They spend much of their scientific lives peering into the deep recesses of space, hoping to chronicle our universe, which took shape some fourteen billion years ago amid a tiny pinprick of starlight and a Big Bang. I'm not in the same league as the astrophysicists, of course, but camping at night inside the inflection of Haleakala, looking up at the sky, I thought I, too, might well have been witnessing the Dawn of Creation.

Another oceanic star-gazing incident found me afloat in a dugout canoe in a lagoon among the chain of coral atolls that make up the tiny country of Tokelau. The islets are situated midway between Hawaii and New Zealand. To reach them, I traveled aboard a cargo ship on its month-long voyage from Samoa. I slept below deck in sweltering quarters reeking of fuel and rat feces and ate foul-tasting food in the seamen's canteen. The surly, tattooed crew looked as if they'd been discharged from an island penal colony. The cargo was diesel fuel and beer. Both were in high demand in Tokelau: measured per capita, it ranks as the top beer-drinking country in the world. The seamen raided the vessel's consignment each night during our long voyage, drank heavily, and fought with one another over trivial matters. One night we hit a reef while the skipper was passed out from drinking whiskey and his beer-sodden crew distracted by a card game. The chief engineer was sober enough to back the ship off the rock with a hard, shuddering reverse thrust of its engines. We escaped with only minor scrapes to the hull. I learned later that this same vessel and crew suffered a more serious wreck on a reef in the Fijian Sea, which punctured the ship's hull and left the crew stranded on the sinking vessel for hours until a navy boat came steaming to their rescue.

Upon arrival in Tokelau, I found no hotel or restaurant because no tourists visit the place. It is too far off the usual paths. I stayed with a family descended from the island's settlers. My hosts related their family's genealogy in bits and pieces at mealtimes and showed me numerous places that marked their island's history: the place where the Polynesians first made landfall, a curious star-mound once used for rituals, fishing lagoons and a rock-enclosed aquaculture pond, copra plantations located on islets across the lagoon, the burial island where family graves are hacked out of the coral reef, a memorial stone chiseled with the dates of a succession of storms and ocean waves that have swept ashore with increasing frequency and velocity. The marker provided tangible evidence of changes in the planet's environment that directly threaten the tiny atolls. Tokelau has the dubious distinction of being one of the first countries that will become totally submerged by rising seas associated with climate change.

My host's forebears lived self-sufficiently on the island. They fished in the sea and cultivated tiny plots of taro and fruit trees in raised beds filled with organic debris. Tokelau had proved such a flimsy foothold for the first Polynesians heading into the Pacific Ocean that it never was a densely settled place. The atolls and lagoons are all that remain of ancient volcanoes that had once occupied the spot. Those fiery mountains, anchored to the drifting Pacific plate, eroded long ago, leaving behind only an encircling trace of bead-like coral islets set within a great arc of a lapis ocean. Tokelau's lagoons are some of the finest bodies of water in the Pacific Ocean: crystal clear and palm-fringed, brimming with fish and coral. On calm sunny days, their glassy surfaces shone like the nacre of a pearl oyster. Adrift at night in one of the lagoons, floating in the tectonic shadow of a long-vanquished mountain and at a great distance from Maui's Haleakala Crater, I watched a galaxy of stars that was conspired by a single celestial event and shared by the two places. The rent of oceanic space was sutured by the milky stream of starlight above me, by the transoceanic passages of the ancient Polynesian navigators, and finally, by my own gaping stare into the incomprehensible yawn of the Universe.

The Polynesian voyages across the Pacific Ocean required the seafarers to know it well. Their knowledge was passed down the generations in song, chant, and dance — the cultural traits that still bolster a Polynesian way of life. When modern-day Hawaiians dance the hula among themselves, it is an ancient sacrament they are performing and not the theatrical spectacle seen by tourists in a Waikiki nightclub. The hula movements evoke trade breezes and ocean swells, currents and tides, island topographies, and thus choreograph the mental maps that guided the Polynesians on their ocean journeys. It's not just that the navigators learned the names of individual stars in order to track their voyages. They knew constellations and star trajectories, which changed day to day, and predicted where a specific star would rise or set on the horizon on different days in a year. They created in their minds a star compass. Here, they fitted distant islands, anchoring them to maps in their minds, and from this ancient cosmological chart of the world, they set sail across the equatorial Pacific Ocean.

A deep knowledge about stars escaped me as I floated in the Tokelau lagoon, nor could I have located my campsite in Haleakala in a celestial chart of Polynesia. At home in Wolf Gap Holler, seated beside my goldfish pond at night, enjoying the *son et lumière* show, land-locked, I have no need of oceanic navigation. I already know where I am and it's not on a voyage across the South Pacific. I'm sitting on a bench in my garden at home. My life in Wolf Gap Holler doesn't depend upon such finely-tuned star reckoning. I'm not certain that my fate is determined by the zodiac, and so I don't feel compelled to identify constellations in the night sky. I know the Big Dipper, Orion's Belt, and the North Star because they're obvious; not to recognize them would border onto being negligent. The stars I see above my goldfish pond are stellar players in my garden's evening entertainment, a grand theater of nature, but I don't know them well at all.

Still, I wonder about the galaxy whenever I watch it progress through the apogee above me. While I neither see my future hitched to the stars nor read a horoscope, when I gaze at the glittering night sky I'm always struck by the notion of infinitude and how it plays out in my life, by the timeless, unbounded, and finally, incomprehensible force of life, and this, in turn, compels me to gaze inward and to reflect anew on my place in the world, to imagine again the course of my life, tracking backward and projecting forward, and so I suppose one could say, because of this sort of self-reflection prompted by the sight of the stars, this inward-turning and -tuning, and the sorts of realizations and even decisions that may come of it, that my future, indeed, is hitched to the stars, just as surely as my carbon-based body derives from the dust of stars. I keep a copy of a Japanese poem in my office at work which has something to say about these things. Here it is in its entirety:

A Love Letter by Nanao Sakaki (1976)

Within a circle of one meter
You sit, pray and sing. Within a shelter ten meters large
You sleep well, rain sounds a lullaby. Within a field a hundred meters large
Raise rice and goats. Within a valley a thousand meters large
Gather firewood, water, wild vegetables, and Amanitas. Within a forest ten kilometers large
Play with raccoons, hawks
Poisonous snakes and butterflies. Mountainous country Shinano
A hundred kilometers large
Where someone lives leisurely, they say.

Within a circle ten thousand kilometers large
Go to see the southern coral reef in summer
Or winter drifting ice in the Sea of Okhotsk. Within a circle ten thousand kilometers large
Walk somewhere on the earth. Within a circle one hundred thousand kilometers large
Swimming in the sea of shooting stars. Within a circle a million kilometers large
Upon the spaced-out yellow mustard blossoms
The moon in the east, the sun west. Within a circle ten billion kilometers large
The Galaxy full blooming in spring. Within a circle one billion light years large
Andromeda is melting away into snowing cherry flowers. Now within a circle ten billion light years large
All thoughts of time, space are burnt away
There again you sit, pray and sing
You sit, pray and sing. ❖

– Lesson Three: The Social Life of an Aquatic Plant –

Plants have an indispensable place in a goldfish pond, as they do elsewhere in the garden. Without them, a pond is a handy reservoir, useful for raising Koi or trout hatchlings, but it's not an ecosystem so much as a fish-filled cistern. And a garden without any plants is ... well, not much of a garden, unless one has in mind a Zen garden like one finds at the famous Ryoanji Temple in Japan, which consists of rocks and raked gravel and illimitable space for introspection. Earlier we learned that aquatic plants are essential to the health of a goldfish pond; when they're present in a pond, all harboring life looks and feels good.

We can introduce some plants to the pond soon after filling it with water, once we've had time to become acquainted with its presence in the garden, noting its scale, relationship to the land, how it fits into the greater design of our home, gardens, and life. A pond can be a centerpiece of a garden, but it should not overwhelm its other features. A goldfish pond is not meant to be large enough for swimming (or fishing), nor small as a bird bath, although birds will occasionally use it for that purpose. A proper fit-to-scale is required for a pond to be the proper garden component.

The selection of aquatic elements is a matter of taste, having much to do with aesthetic inclinations, but I recommend sticking with living plants rather than the fake ones on sale everywhere. Bright plastic flowers jump out at customers in the Big Box garden centers, noticeable even from the parking lot, and considering the number of shelves filled with plastic plants, a lot of people must put them into their garden ponds. The fake plants are shiny and colorful, require little work, and are long-lasting—all popular criteria among consumers. Still, a goldfish pond is alive, and it seems odd to me to want to float a fake plant in it. Would I put a plastic tomato in a bowl of vegetable soup or add a waxed grape to a fresh fruit salad?

Specialty pond stores have the best selection of aquatic plants to choose from, and of course, we can be certain they're alive. Given the variety of plants for sale, though, a person can be easily overwhelmed by the selection. It's helpful in this case to envision the pond in a natural setting, perhaps recalling a farm pond where a person once fished as a kid or a small lake beside which we might have parked the car on a first date, and remember how the water's edge was lined with marsh marigolds or cattails. Swing our mind's eye to the center of the pond, pan its surface for water lilies, fish, and the occasional turtle sunning itself on a partially submerged log. The aquatic plants we would see in such a setting are grouped by botanists into marginal or "bog" plants: water lilies, floating plants, and oxygenators. If we imagine the goldfish pond interfused with just the right combination of these plants, it might be as if it had popped out of the dappled landscapes of our imaginations and landed safely inside our garden.

The marginal water plants thriving at the wet edge of a pond like to stand in damp shallows. If you're just now designing the pond and haven't yet dug a hole for it, consider shaping a small boggy extension at one end to hold marginal plants. A bench for this purpose can be made by creating a shallow peninsula filled with clay-based soil. Make a low retaining wall of stone and mud to allow water to slowly seep from the main pond area to the boggy extension and to saturate the soil. You might be tempted to visit a nearby lake to scoop out a bit of the bottom muck for the pond, but I recommend not doing this as you will likely and inadvertently bring into your garden the seeds or roots of lake plants. Many of these species multiply and spread fast, grow very large, and would quickly exhaust the tiny space of a goldfish pond. Marginal plants should stay at the edge of a pond, where they belong, and not invade its placid interior.

One of the most common bog plants in eastern North America is the marsh marigold (*Caltha palustris*). *Palustris* is Latin for "of the marsh," so we can see right away where this plant belongs. It's a showy member of the buttercup family that is very popular in England where it's also known as horse blob or molly blob. The marsh marigold is sometimes called cowslip where I live, although that name also applies to *Primula veris*, which is a northern cousin of the marsh marigold. I don't grow the plant in my pond because it's too big and bushy for the small scale of it; the marsh marigold can grow a meter high and spread fast given an opportunity. The marsh marigold also is mildly toxic and may cause epidermal rashes among some persons. I've had many bad episodes in the past with poison ivy and tend to shy away from plants that threaten my skin. However, with its bright yellow flowers and thick waxy foliage, I can see how the marsh marigold would make a lovely marginal plant in the right goldfish pond.

I grow three types of bog plants in my Wolf Gap goldfish pond. One is the pickerel rush (*Pontederia cordata*), a species that has tall, spiky stems that overhang the surface of the water. Their purple blossoms attract busy squadrons of iridescent dragonflies. The other two marginal plants in my pond are types of irises: a violet Japanese iris (*Iris enstata*) and yellow flag iris (*Iris pseudacorus*). I hesitate to mention the yellow flag variety; following its introduction to North America from Europe as a garden plant in the early 1900s, the *Iris pseudacorus* spread rapidly to become an invasive weed, especially in the American

South where it has overrun pristine native habitats and displaced indigenous plants. It is banned now in many places because of its obnoxious and bullying behavior toward the native species. I'm not sure if it's okay even to continue to grow it in my pond. With a moniker like "yellow flag," which is the warning signal in America for everything from football games to NASCAR races, I should have known better than to put the plant in my Wolf Gap garden. In my defense, though, I think it would be extremely unlikely for the *Iris pseudacorus* to leap out of my goldfish pond and go running through the woods looking for another wet place to colonize. They just don't do that well outside of a marshy habitat, and a specimen would have to stray far from the garden before coming upon a suitable spot in the forest to set down its roots.

The yellow iris is called a "flag" iris not because it is some kind of pennant but rather because it looks like a rush or a reed prior to inflorescence. The old English word for rush is *flagge* and hence, the plant's etymology. In France, the iris is called the *fleurs de lis* or "lily flower," referring specifically to the *Iris pseudacorus*, and was commonly used as royal insignia on family crests and banners. Some additional confusion is introduced by the fact that prior to the nineteenth century, other varieties of iris were also called a lily elsewhere in Europe, and in some places they still are known that way. It's clear from the historical record, though, that the famous *fleurs de lis* in France refers to the yellow flag iris and not to a lily.

Since the *fleurs de lis* appears on so many royal badges and the plant comes from the dynastic gardens of old Europe, it seems necessary that whenever I consider the social life of the *Iris pseudacorus* growing in my goldfish pond, I acknowledge its botanical role in the imperial history of Western civilization. The iris appears, for instance, on the coat of arms of the King of Spain and the Grand Duke of Luxembourg. Its shape was stamped onto the crown jewels of England and Scotland, and the florin coin of Italy. It was used in medieval Bosnia to denote the ruling House of Kotromonic. The shape of the flower held a prominent place in early Christian art where it was associated with Mary and hence came to symbolize purity and chastity. It appears still among the iron gates of the Vatican fortifications.

Mainly, though, the emblem of the *Iris pseudacorus* is associated with the French monarchy where it first signified the reign of the Frankish king Clovis (c. 466–511). Later, under King Louis IX (1214–1270), the three petals of the flower came to represent faith, wisdom, and chivalry. In the eighteenth century, it appeared on the French flag carried by Great Lakes fur traders exploring North America for beaver, fox, martin, mink, otter, and other valuable fur-bearing animals. It still appears on the flags and wrought-iron fences of the French colonial cities like New Orleans, Vincennes, and St. Louis. It shows up on the logos of many North American sports teams and universities. It even appears on the grills of Chevrolet Corvettes and is affixed to the labels of Campbell's Soup cans, so famously painted by the artist Andy Warhol. It is also the symbol of the World Organization of the Scout. The iris in my garden has none of these ambitions. Its triadic shape, though, heralds the purpose of my goldfish pond: a place for contemplation, wonder, and rest.

The water lily is a useful plant to consider for a goldfish pond. In fact, a pond without a water lily is hard to imagine. Although the lily has been confused with the iris on the epaulette of royalty in Europe, this mix-up mainly applies to the terrestrial varieties; when we come upon a water lily in a garden pond we will notice right away how it differs from an iris. Like the marginal iris, water lilies root in muck on the bottom of a pond, but that's about all the two plants share in common. The latter disperses across a pond with fat, plate-sized pads from which emerge white, saucer-shaped blossoms, and look nothing like an iris, which stands upright in the water with knife-sharp leaves and a forked flower. Despite any lingering botanical incertitude among the royal families of medieval Europe, I find the two plants growing side by side in my pond to be as distinctive from one another as pieces of cutlery in an *al fresco* garden luncheon.

Water lilies are especially valuable in a goldfish pond because their floating pads cover much of its surface, reducing the amount of sunlight entering the water, and this, as we learned in lesson two, helps to ensure a translucent pond. Meanwhile, their delicate white blossoms make a gracious sight, and we know from Bali that this kind of ornamental beauty helps keep the world in balance. When we put a water lily in a goldfish pond, we not only enhance the clarity of the water, rendering a healthier and more attractive pond, but also, because of its great beauty, we ever so slightly increase the odds of human life on Earth achieving a sense of equilibrium. Lilies jump out of their pots, though, and may scoot their rhizomes across a pond bottom faster than a fleet of windsurfers on the Hood River in Oregon. Caution and attention thus are necessary to keep a check on lilies once we've introduced them into the pond. They've surpassed their allowable space once their fat leaves no longer float raft-like on the water's surface, but rather arch toward the heavens as if petitioning a place in the sky itself.

The seventy common species of water lily are divided into two main categories: hardy and tropical. Depending upon climate, we have few or many options. The hardy varieties effloresce only in the daytime, while the tropical ones bloom day and night. The tropical blossoms come in numerous hues, whereas the temperate varieties generally have only white petals with a hint of yellow showing at the stamen. If a person wants a blue water lily blooming at night in the goldfish pond, they need to live in a warm climate—or go with a plastic one, which isn't recommended for the reasons noted above.

The peltate leaves of a tropical lily can be quite large: three meters in diameter in the extreme case of the *Victoria amazonica* found in Brazil. A single pad from this South American giant would cover the entire surface of my goldfish pond in Wolf Gap Holler. Otherwise, the variations among the temperate lily species are minor: floating leaves of different size or subtle differences in the hue of a bloom. One might invest in tropical lilies for a temperate pond, but then they would need to overwinter indoors, which can be a challenge because they also require fulltime immersion in water. All lilies, regardless of variety and climatic inclination, effuse tethered pads that float on the surface of a pond. They have at least that much in common with members of the next category of aquatic flora: floating plants.

Like lilies, floating plants are also invasive, so you need to be wary of them. Water lettuce and hyacinth are manageable at the scale of a goldfish pond, but I'd keep away entirely from fast-proliferating duckweed. If you put duckweed in the water, there may come a day when you awake in the morning, head to the goldfish pond with a cup of coffee in hand, and find that you can walk across it. It wouldn't be a miraculous occurrence, not like Moses parting the Red Sea or Jesus of Nazareth walking on water, just a soupy, green mess so thick with duckweed that it might hold a person's full weight.

When selecting floating plants, it's best to stick to a local gardening store and not go traipsing off to a nearby lake to obtain floating plant to drop into your pond. We're asking for trouble if we do that. Errant hyacinths have shut down entire shipping lanes on rivers in China and Brazil, and their multiplying masses clog lakes in southern India to the point where the water bodies are no longer navigable. Floating plants are valuable ecological additions to a pond because they partner with other plants to block sunlight and provide anchorage to algae during the minor blooms that can occur, but we have to ride herd on these creatures like a cowboy on cattle grazing on the open range. Otherwise, they'll grab more than their fair share of the pond's sunny surface.

Oxygenating plants are a final type of aquatic flora to consider for the goldfish pond. They live entirely submerged in water as denizens of the garden lough, where they release oxygen and compete with algae for nutrients, thus working their magic to help maintain clear water in a pond. The common varieties of oxygenators include monkey tail, hornwort, and parrot's feathers. These plants may be familiar because they're the specimens we commonly see growing in fish aquariums. Unlike the perennial iris and water lily, oxygenators are not cold hardy and need to be replaced each growing season. The easiest way to keep these plants in a pond is to bundle them with a rubber band and anchor the clump to the bottom with a weight.

In the vast catalogue of aquatic plants, one stands out in my goldfish pond among all others: the sacred lotus (*Nelumbo nucifera*). It rises like a pink queen amid the raft of water lilies, asserting its regality over them and diminishing the lily's pure white flower to the role of admiring courtier. I've been enthralled with the lotus ever since I first saw them blooming in the mountain lakes of the Vale of Kashmir, a place described by the Moghul emperor Jahangir (1569–1627) as paradise on Earth. The Indian emperor was so moved by the sight of them that he imported the lotus to grace his own palatial gardens and deployed its motif as the centerpiece of murals that cover mosques and mausoleums located throughout the Islamic world. It is carved into countless Hindu sculptures.

The lotus also is the sacred flower of Buddhists, for whom it symbolizes purity, compassion, enlightenment, and love. The Tang Dynasty poets in China exalted the lotus, a flower that emerges out of the muck of a pond to become a thing of wonder and beauty. The plant came to symbolize for them a disciplined person attaining eminence from obscure and humble beginnings. Its blossoms, plucked fresh from a pond, are offered at temples and shrines throughout Thailand, Laos, and Bali. Perhaps most famous of all, the Buddhist mantra *Om mani padme hum* invokes the lotus in a refrain that is widely translated as "Hail, the jewel in the lotus." This prayer is chanted by monks, carved into rock tablets, and impressed upon the spiritual imagination of people throughout the world.

The sages all agree: a lotus in a goldfish pond is like a flash of insight, a brilliance that can fill an empty mind. I know it always is a lovely sight in my Wolf Gap garden. Chinese philosophers believe that once someone sees a lotus in a pond, they are never again the same; it is that pure and life-changing. In Egyptian mythology, it represents the sun god rising from darkness to illuminate the world. Hinduism associates its efflorescence with an awakening consciousness. In Cambodia the lotus is used as an herbal medicine, and in Burma (Myanmar) it is dried and pounded into incense. An important shrine in a Balinese temple is the Padmasana, or Lotus Throne. This sculpted stone seat supports a benevolent deity and significantly faces away from the violent volcano Gunung Agung, as if turning its back in silent disdain for the destruction wrought by this lesser mountain god. Many of us know Padmasana as a popular sitting posture in various schools of yoga and meditation.

Nowhere, though, have I come across the lotus more regally presented than among the ancient ruins of Angkor Wat in present-day Cambodia. It still blooms profusely in the moats surrounding the great stone monuments, and is carved into countless bas-relief galleries, monoliths, and statuary. Angkor is the site of what was a vast and productive hydraulic civilization under the rule of Khmer kings; water was harnessed to irrigate rice fields that supported more than a million residents, and the lotus blossom signified fecundity in an empire that had no worldly peer. Present-day ruins mark Angkor as the world's largest religious site, dedicated first to Hinduism in the ninth century and then to Buddhism in the last quarter of the twelfth century, lorded over by men who believed they ruled by God's decree. The lotus was their sacred motif and served as the totem of a celestial kingdom. It appears throughout Angkor as a metaphorical lineament carved into stone, such that its divine countenance might forever grace the face of Earth.

The golden age of Angkor dates from 900 C.E. to 1200 C.E, when the complex sprawled across a thousand square kilometers of jungle and rice farms. It contained countless temples, towers, embattlements, and ceremonial plazas. Most of it long ago disappeared beneath a thick cloak of tropical plants. The jungle-wrapped stone structures of the ancient civilization now lie scattered about the landscape like the playhouses of god-kings, the verdure armature of a powerful kingdom that eventually was overcome by rot and hubris from within, and then, outwardly, by centuries of equatorial decay so that its exoskeleton appears now as a crumbling and magnificent shell strangled by ficus trees and flowering vines. This lush decay evokes the megalomania of its famous builders: the Khmer kings Jayavarman VII (1125 – 1200), whose monolithic stone head overlooks the Angkor Thom site; and Suryavarman II (1113 – 1150), the famed creator of Angkor's sacred temple. Both men deigned themselves divine rulers and anchored their civilization in a supernatural reckoning.

The most prominent component of Angkor is the central sanctum, a collection of upward soaring ramparts shaped in a spiraling mandala that represents the mythological mountain Meru, home of gods, center of the world, axis mundi bridging Earth and Empyrean. The Khmer kings had centered their reign in this imagined universe replete with cosmological orientations: stone monuments in conformation with the epicycles of the sun and lunar orbits, observatories from which priests once followed the paths of the planets, and multi-tiered pavilions ascending as stairways to heaven, all overlooked by the soaring adamantine visages of the god-kings. Most striking of its features are the five sandstone towers located at the exact sacred center. Each is shaped in the likeness of an unopened bud of a Lotus flower. They stand sixty-five meters tall and represent the five celestial summits of Mt. Meru.

The crenelated lotus bud towers of Angkor puncture the steamy green horizon and pierce the sky in a famed silhouette that arises from an encircling moat of pink and yellow lotus blossoms. Around the lotus towers are the geo-selenic temples, famous ones such as Bayon, Preah Khan, and Ta Prohm, as well as numerous lesser shrines, all pieced together as an arcane sacred geography meant to align the Khmer kingdom to heavenly bodies and thus to bolster the kings' claims to a divine right of rule. Standing high above everything else — the ponds and canals, pediments, gilded stucco towers, bas-relief, and slaked lime walls, even the carved heads of the kings — is the quincunx of lotus bud towers. They soar skyward as if making silent supplication to the Khmer gods, a grandiloquent gesture of praise carved into stone so that it might last forever.

I visited Angkor in September, when the countryside is saturated by water from the monsoon rains, and I was dazzled by the presence of water everywhere in the landscape: shimmering ponds and lakes reflected the white-hot sky, rivers coursed through emerald fields of rice, a seemingly eternal mist rose in smoky plumes from the dark tangles of jungle. The floodwaters lent an otherworldly appearance to the place. Eponymous temples, long ago abandoned and now crumbling amid the heat, humidity, tree roots, and vines, floated amid their diluent moorings as puzzling fragments of a long-lost Leviathan civilization. Their edifices multiplied in reflections cast by the still pools of water, fracturing amid surface ripples to render a pixelated tableau, as if a digital stage created for yet another archaeological thriller. But it was all quite real and, I thought, a proper setting for the preternatural lotus flower.

Given such eminence, how could I not put a lotus in my goldfish pond? Its perfectly formed, super-hydrophobic leaves shed raindrops like tiny pewter balls rolling quicksilver off the plant to splash into my pond, stirring its surface with tiny ripples that flow across an entire civilization. Its delicate pink blossoms set on tall, thin stalks angle ever so elegantly toward the sun, slowly rotating to follow the solar path of a summer day. Like a rare, black-necked crane peering intently upwards at a frosty glacier glistening somewhere in the high Himalaya, the lotus in my Wolf Gap garden appears to be examining its rightful place in the world. With an imperious bearing, knowing that its roots lie far beyond the topographical confines of my tiny Kentucky valley, the lotus growing out of the muck in a pot I set into my goldfish pond connects me more than anything else in the garden to the distant wonders of the world. ❖

CHAPTER THIRTEEN

I can sit in my garden and observe its natural elements, understanding that I am more than audience; I am part of them and so bear witness to how beauty and balance is all around me, anchoring me to a prospect of an harmonious life, and considering how best to conduct my life amid global trends that are shaping our planet. Like a mirror set into the glebe of Earth, my goldfish pond stares back and prods me to consider once again my place in the world.

A Point of View

Across a scallop-shaped bay from the Aegean port of Parikia was a pocket beach protected by shallow dunes and shaded by the canopy of a grove of resin-scented mastic trees. It looked to be a perfect place to camp beside the sea. I had disembarked my vessel earlier that day when it stopped at Paros Island to offload supplies and wouldn't resume my journey for several weeks. The lengthy stopover wasn't my original plan. I left Athens intending to hopscotch across the Aegean Sea on local ferry boats, halting at a spot just long enough to see some of the local sights before booking passage to another islet and thereby take in a good part of the Greek archipelago. When my ship pulled away from its dock, though, I remained behind on Paros because I thought I might see more of the Mediterranean staying put in one place than skipping across a bunch of islands. I also was feeling lazy, having traveled hard for the past few months, and Paros looked to be a good place to stop and relax. So I shouldered my backpack and made my way round the harbor, setting my tent into beach sand under the shade of a thorn grove beside sparkling blue water.

My days on Paros were spent in much the same indolent way. I swam across the bay in the morning when the sun was still low in the sky, dined at a village café, drank cheap wine called Retsina, which has been made on the island for at least two thousand years, played cards with the local layabouts—a game called *bastra*—rambled around the village or napped in a shady spot during the afternoon, and finally returned to my campsite at dusk. It was a pleasant way to pass the time. All the while, the hot sun rose, arced across the sky, and sank back into the sea in a lazy pendulum that perfectly matched the timings of my own lethargic perambulations. The Homeric concept of *arête* often is applied to characteristics of human excellence, but I thought it might just as well be a paean to an ordinary day spent on a sunny Aegean islet. Surrounded by Cerulean waters, I watched each day on Paros slip lightly into another, and when evening approached I was back at my campsite, ready to sit among the stars with bioluminescent waves rushing toward me on the beach.

Paros is a member of the Cyclades, a group of two hundred islands of varying size scattered in the sea north of Crete. Unlike its trend-setting cousins Mykonos and Santorini, or the sacred island of Delos, which attracts throngs of pilgrims, Paros at the time I visited wasn't a popular landing spot for tourists. It had quaint, white-washed homes colored by sprays of bougainvillea, scenic little vineyards dotting the countryside, and magnificent vistas, but not much in the way of clubs or nightlife. The small hamlets lacked shopping boutiques, tourist services, or the prominent historical monuments found elsewhere in the Mediterranean region. It wasn't a port of call for big passenger cruise ships. I remember it as a lovely but overlooked place. Sleepy and nondescript, Paros also proved a turning point for me, the first in a line of stepping stones, which, like platforms stretching to the horizon, beckoned me toward a life I had not yet imagined.

I had been intending to return to Michigan at the end of my summer backpacking tour of Europe and continue with my college studies. I aborted that plan on Paros and decided to continue my journey instead. Sharing stories with other travelers, some of whom had spent many months in Asia, I wondered at their tales about places located further east: souks and mineral baths in Constantinople, crumbling mud mosques in Khorasan, camel caravans crossing the Hindu Kush range in Afghanistan, and tiny Buddhist villages nestled amid the stony ramparts of the Himalaya. Such imagery captivated me, and as the idea of visiting Asia took shape in my mind, the prospect of returning home grew distant.

I decided then not to retreat westward but to continue further east. Like Paros' stone-buttressed windmills lazily turning in an afternoon breeze, I found myself propelled by new winds that had shifted direction. I slowly pirouetted on the island. With an Oriental prospect now on the horizon, I embarked on an overland expedition first to Istanbul, where I crossed the Bosporus Strait from Europe into Asia, and then across the empty, bandit-strewn wastes of eastern Turkey to the deserts of Iran. It was a mad dash across lands that today are in a tragic state of turmoil. I halted in Afghanistan, resting in its capital city of Kabul, where I found budget lodging on Chicken Street and ate simple meals of fire-roasted goat and flatbread served piping hot from clay ovens. I learned about Sufism and the mystical poems of Rumi, was held at gun-point by two rough Pashtun men in scraggly beards and turbans, smoked opium and hashish, all the while pushing hard against my life's past conventions. I heard about more wondrous places and went to see them: giant Buddha statues carved centuries ago into the cliffs of Bamiyan Valley and not yet blown apart by Taliban missiles, travertine lakes of Band-e-Amir, adobe caravansaries straddling the Silk Road. Finally, I descended the Khyber Pass into Pakistan and crossed into India, where I left behind the mosques, dust, and ululations of Central Asia and entered Shiva's colorful world of Hindu rituals and raucous bazaars. The crowds and heat of the Indian towns became insufferable, and I sought escape by traveling to the Buddhist valleys located high among the snowfields of the

Himalaya. There I found magical places of wonder and adventure, which over the course of my life have become vital coordinates in my world.

At the close of that travel year, I returned to the American Midwest to finish my college studies in business. But the trajectory of my life had shifted in a major way. Back on campus, I walked into a large auditorium filled with hundreds of students as a guest speaker extolled the virtues of a career in corporate finance. I was out of place, vertiginous, and when my discomfort intensified I gathered my things and stumbled in the dark past the rows of students to the exit in back. As the door closed upon the lecture hall, I knew I had abandoned not only the speaker and the study of business, but that kind of life. I felt a huge wave of relief—escape—washing over me, and then, once I had gained the out-of-doors, a disquieting inquiry: what now?

The question has become a driving force in my life, fueling adventures that take me from my goldfish pond in Wolf Gap Holler to distant places on the planet. It prompts me to think daily about old ideas and to constantly entertain new ones. I sometimes recount that long-ago day of reckoning on the Business School campus with my own students, who often worry about what the future will bring them after they finish their college studies. We discuss how a change in the way a person thinks about the world might lead to fundamentally new ways of seeing it. I liken this shift in apprehension to the ideas of theologians who refer to moments of epiphany and of psychologists who label them as gestalt. In academic parlance, we speak of a paradigm shift: grasping a new worldview that tugs at a person and asks one to think differently, to see the planet and the course of life in novel ways. I find the talk of world travel in my geography classes to be a perfect ploy for touching upon such things with my students. Otherwise, they are delicately held matters. We discuss places we'd like to visit and why, and consider the journey that might be required to reach them, a journey that is as much inward as any campaign to see exotic new destinations and thus as likely to occur to a person seated thoughtfully beside a goldfish pond as to someone out globetrotting.

The improbable intertwining of travel and a garden has taken form in my life as one begets the other. I look upon my goldfish pond and see in it a world of stillness in motion. The work of gardening, itself, is a kind of moving meditation for many people. To find one's center in a garden is a kind of journey. T. S. Eliot expressed it well when he wrote, "And the end of all our exploring will be to arrive where we started, and know the place for the first time." The very idea of travel without motion is suggested by any number of metaphysical concepts: astral projection, out-of-body experience, spirit walking, or time travel. It is found, too, in the teachings of widely practiced religions: Paul's *Second Epistle to the Corinthians* describes astral projection among Christ seekers; the Quran speaks of "night journeys" when the soul is traveling while the body is asleep; Egyptian Coptic teachings describe the spirit as an entity hovering just outside a person; Indian yogis and Tibetan monks claim the ability to levitate and travel to distant holy places. Most of us, in fact, journey in our minds to distant locales without ever taking a stride. We call it daydreaming. This most ordinary of human pursuits can lead us to the very frontiers of our geographical imagination, and thus onto new outlooks on life. For the record, I have discovered few places in the world to be better suited to daydreaming than the garden bench located beside my goldfish pond in Wolf Gap Holler.

Still, a life of gardens and travel may seem incongruous, as if the one could not support the other, when in fact the two are reconciled even by etymology: the word "garden" comes from the old Middle English term "gardin," meaning enclosure, while travel derives from the French word *travail*, meaning to labor or toil. When we see the world as a garden, our travel becomes a labor of love and exploration. The terms become further intertwined when life is understood as a journey of discovery and hard work that leads inward to a sense of wonder and purpose.

It is profoundly challenging, though, to transfer such lessons to ordinary days in modern times. We live in a multi-tasking society and easily lose focus. It seems to require the skills of a juggler. Multiple events close in upon us with a quickness that takes the breath away. Yet, we also mistake multi-tasking for juggling, when the two could not be further apart. Juggling enlists concentration and prescience that is difficult to achieve in the scattershot approach of so much of our information-laden modern life. No wonder people are dropping the ball so much of the time. We have become distracted.

Many years ago when I began meeting my Wolf Gap neighbor George in the morning for coffee at my goldfish pond, it was in the simple notion that such moments might be pleasurable for us and provide a chance to disengage from societal forces that lie beyond our control. We sit on a wooden bench beneath a magnolia tree and watch the pond spring to life. The fish always appear agitated at our arrival—not frightened, just anxious to be fed. The frogs, meanwhile, peer at us like Zen monks from their solitary perches around the pond, dispassionate and unblinking, as if in meditation among the frenzied school of fish. Scattered here and there amid the pads of greenery are pure-white lily flowers. George and I sometimes use the efflorescence of a water lily as a sort of timepiece to gauge the length of our stay at the pond; it takes about an hour for a blossom to fully unfold once it begins to feel the sun settling upon it, which is also the length of time required for me to watch a sunset at the end of a day, staying to witness the whole glorious alchemy of refracted colors transmuting in the crucible of the evening sky. This is to say, from a certain perspective, the speed of a lily blossom unfolding in a goldfish pond is synchronous with the spin of the earth on its axis.

My pond is animated with insects once the morning air is sufficiently warm. They gyrate and spin above the water like aerial gymnasts. Dragonflies swoop in tight little formations before the pollen promise of aquatic plants and then suddenly stop and hover like trapeze artists while they inspect the nourishing anther of a flower. Ruby-throated hummingbirds dart in loud and nervous blurs, switching directions on a dime, balancing as delicate, spinning iridescent baubles on a point of air. Fat carpenter bees in yellow-and-black uniforms float above the pond like the Goodyear Blimp on Super Bowl Sunday. Water bugs skip across the surface of the pond and cast redoubled shadows on the rocks below. They are cartoon-like in their submerged silhouette, a confluence of shadow and self that might pose an existential dilemma were they not of the Anthropoda family. Butterflies jostle for space in the air, while frogs leap from lily pads to rocks and fish swim through the clear water. At the height of action, the pond is thick with the ecosystem antics of Kingdom Animalia; it's a regular Barnum and Bailey Circus once the sun settles onto the pond's surface.

All the while, buzzing insects and croaking frogs serenade us with their songs in a choreography of improvised music. Every so often, one of the larger frogs goes off on a baritone riff like a Tuvan throat singer stepping up to a microphone: бир—ийи - Khynaar - бир—ийи (One Two Testing: One Two). It's sometimes hard for George and me to carry on a conversation amid all the ruckus, and we're forced, instead, to quietly sit and marvel at the virtuosity of the dragonflies or the syncopation of the amphibian chorus. It's customary for George and me to sit by the goldfish pond on nice mornings when we have the chance. Such meetings help to hold our worlds together and have become a small part of the pattern of our lives in Wolf Gap Holler.

Some mementos I've collected on my distant travels make their way home and into my garden. Odd little souvenirs—carved talismans, spirit houses, prayer flags, and temple bells—embellish my goldfish pond, piling up hard-earned memories of distant travel and giving the place a cosmopolitan personality. These artifacts are conveniences for me. Seated among them, I am easily reminded of lessons I've learned in places I've been and tasked now to apply to my life at home. I can't levitate like a Tibetan monk and whisk myself away at a moment's notice to a monastery in the lap of the Himalaya, or hop aboard an outrigger canoe and navigate by the stars across the Pacific Ocean, but I can sit in my garden and observe its natural elements, understanding that I am more than audience; I am part of them and so bear witness to how beauty and balance is all around me, anchoring me to a prospect of an harmonious life, and considering how best to conduct my life amid global trends that are shaping our planet. Like a mirror set into the glebe of Earth, my goldfish pond stares back and prods me to consider once again my place in the world. ❖

About the Author

David Zurick travels often from his Wolf Gap home to photograph and write about distant places and cultures. He is the author of numerous books and maintains a photo blog called "Picture Pilgrim." You can find out more at: www.davidzurick.com.

CPSIA information can be obtained
at www.ICGtesting.com
Printed in the USA
LVOW10s0535150917
548811LV00020B/1565/P

9 781947 067028